SURVIVING TO DRIVE

SURVIVING TO DRIVE

A Year Inside Formula 1

GUENTHER STEINER

TEN SPEED PRESS
California | New York

First published in Great Britain in 2023 by Bantam, an imprint of Transworld Publishers

Typeface: Typeset in 12/17.5 pt Bembo Book MT Pro by Jouve (UK), Milton Keynes

Library of Congress Cataloging-in-Publication Data is on file with the publisher.

Hardcover ISBN: 978-0-593-83547-0
eBook ISBN: 978-0-593-83548-7

Printed in the United States of America

Editor: Aubrey Martinson | Production editor: Serena Wang
Cover design: Anthony Maddock/TW | Cover photography © Team Haas
Art director: Yang Kim | Production designer: Mari Gill
Production manager: Dan Myers

10 9 8 7 6 5 4 3 2 1

First US Edition

'A lot of people criticize Formula 1 as an unnecessary risk.
But what would life be like if we only did what is necessary?'

– Niki Lauda

CONTENTS

Guenther Steiner is without any doubt one of the most unique human beings I have ever had the pleasure of meeting. Not just in motorsport, but in all walks of life. Searingly honest, dangerously forthright, infuriatingly stubborn sometimes, unusually reliable, occasionally inspirational, constantly amusing and, unless there are children in the room, mainly unfiltered! Even the way that Guenther talks is unique. After all, do you know anybody else who sounds like Guenther Steiner? I don't.

I first met Guenther twenty-two years ago while he was working for Niki Lauda at the Jaguar Formula 1 team. I remember thinking to myself after meeting him, *Wow, where on earth has this guy come from?* When I learned that Guenther had been born and brought up in the Italian city of Merano on the Italian–Austrian border, and that he had been working in rally for several years prior to joining Niki, everything made sense. People who work in rally rarely mince their words, nor do people who originate from that part of Italy. He was and still is very much a product of both his people and his surroundings.

It was only when Guenther approached me at Ferrari many years later that I began to get to know him. He had a dream of creating an American Formula 1 team and wanted my help. Guenther will tell you the story himself, but what I want to say here is that his passion for the project and his knowledge of Formula 1 and of motorsport were as integral to me wanting to support his dream as the concept itself. That concept, of course, became the Haas F1 team and although I am obviously partisan in my position as CEO of the Formula 1 Group, I am proud of the part I played in bringing the team to life.

After a business relationship was formed, Guenther and I were able to develop a friendship – a friendship that has evolved over the years and which I think we both value enormously.

A question I am sometimes asked is whether or not I had any idea that *Drive to Survive* would catapult my friend from being a generally anonymous but well-respected team principal into a superstar. The truth is that, as much as I value Guenther as a friend, I never considered for a moment what effect his personality and character might have on the general public. Or on the sport, for that matter. Had I known, I might have warned everyone what to expect! The fact is, though, that the show has created a monster. A monster who appears to have charmed half the world and who I believe is a genuine force for good in our sport.

Once it became apparent that Guenther had caught the audience's imagination I gave him one piece of advice, which was to be sure that you stay as you are. 'Never change, Guenther,' I said. 'Always be you.' Fortunately, I think he always will.

I hope you enjoy the book.

Stefano Domenicali
London, January 2023

OFF SEASON

Monday, 13 December 2021 – Yas Marina Circuit, Yas Island, Abu Dhabi

It probably won't surprise many people that I am starting my book with a swear word but all I can say is, thank fok that season is over! It's been a nightmare from start to finish. I don't drink very much but this year I've been tempted to take it up professionally. Whisky on a foking drip. That's what I needed sometimes!

It hasn't just been this year, though. The shit goes back even further than that. Being kicked out of Melbourne at the start of 2020 was probably where it all began. We thought we'd be racing again in a couple of weeks' time but what was actually ahead of us all was months and months of uncertainty. Will we survive? Will we even race again? Nobody knew, you know. It's no secret that there were probably four teams that could easily have gone under during that time, including us. Pete Crolla, our team manager, had meetings with the FIA and Formula 1 two or three times a week, and while he fed everything back to Gene Haas and me, we tried to keep everything afloat. Even the sport itself was under threat for a time because we didn't know how long the pandemic would last. Would it be three months? Would it be three years? Would it be three generations?

In the end Formula 1 basically shut down for about ninety days. That's incredible, when you think about it. Especially for a sport that is famous for progression. The only time this sport ever comes close to shutting down is during the summer break and over Christmas. But even then there is a lot ticking along in the background. Me, for instance. You think that I shut down in the summer and over Christmas? Don't be ridiculous. I have things to do. That ninety days, though, was a pretty shit time.

The thing that Formula 1 as a whole did right during those ninety days was to carry on as if things would eventually get better, at least as much as it was possible. This meant that when we could come up for air and start racing again we'd be ready to go. A lot of people worked hard to make that happen and it was a big risk. I mean, how long can you keep an engine ticking over before it finally runs out of gas or goes wrong? It was a nervous time.

We as a team had to do a lot of restructuring in order to keep us ticking over, so it wasn't just a case of carrying on and hoping for the best. For sure, nobody was able to do that. One element of the 'return to racing' programme that was devised by the FIA and Formula 1 was the continuation of the existing regulations so, rather than produce a brand new car concept for the following season, we had to develop the existing cars. Unfortunately, for reasons I'll very soon explain, our 2020 car wasn't a great one so instead of trying to develop that for the rest of 2020 and throughout 2021, which would have been like trying to polish a turd, to be honest with you, we made the decision to use it again as it was, give or take, and put everything into developing a new car concept using the new regulations that were coming into place.

I have to pay tribute to Gene here because he could easily have taken a different viewpoint and said, *Fok this for a game of soldiers.* A lot of people would have, I think. Especially with all the uncertainty that was still surrounding the sport. Even when we started racing again nobody knew how long it would last. Every day we were reading about new strains of Covid and so we were always looking over our shoulders.

In all my years in motorsport, the decision to write off the 2021 season is the hardest I have ever been involved in. We're all competitive people and so actually choosing to be shit for an entire season goes against everything we believe in and are striving every day to achieve. Every single race weekend put the team on a downward spiral. On arriving at the track they'd all try and be upbeat but then over the course of the weekend they'd start to sink. 'What are we even doing here?' they'd say. 'This is shit!' My main job during the whole of that season was to tell the team as many times as I needed to exactly why we were doing what we were doing and remind them that there was light at the end of the tunnel. Or should I say, the wind tunnel. You see, I'm also a foking comedian!

'Look, guys, better times are ahead,' I kept saying to them. 'You have to believe that.' Fortunately, they did and they stuck with it. We've got a really, really good bunch of people in the team at the moment. Sixty per cent of our staff have been with us for four or five years or more, which is pretty foking cool. We might not be able to retain form very well sometimes, but we're great at retaining staff.

Writing off the 2021 season was the correct thing to do, though. I'm sure of it. Gene is, too. In 2020, which was a normal season for us in terms of spending and development (but turned out to be shit for all kinds of different reasons), our entire budget was roughly

$173 million, whereas Ferrari's was $463 million and Mercedes' even more than that. Almost half a billion. That's a big gap, you know. Even if we'd spent half of our allocated wind tunnel time on the 2021 car we'd still have finished last. Why would we do that? I've been in motorsport for thirty-six years and sometimes you have to just surrender to the circumstances and improve things when you can.

In 2021, as part of the new rules which were designed to make the sport more competitive, the share of the budget that's performance-critical – design and development, component manufacture and testing – was limited to $145 million per team, and in order to take full advantage of that we decided to run the 2020 car in 2021 and put as much money as we could into developing the car for 2022. The top three salaries at each team aren't included in this so it means that the guys at Mercedes, Ferrari and Red Bull can still gain an advantage by hiring the best people. Or at least three of them. I'm OK with that. It's better than it was.

We're all competitive people and every person who works for every team on the grid obviously wants their team to do well. OK, so we're not likely to win many races anytime soon. In 2018, though, Haas, who were, and still are, the smallest team on the grid, scored ninety-three points and finished fifth in the Constructors' Championship. That's not bad for a team that was also only three years old at the time. We're not stupid.

The only thing that really kept the team going last season was that in the background we'd been developing a car that will hopefully make us competitive again in 2023. So far in our history we've had two promising seasons in 2016 and 2017, a foking brilliant season in 2018, a pretty difficult season in 2019, a shit season in 2020 and a dead season in 2021. That's three on each side. There's a hell of a lot riding on what we're trying to achieve right now. Not to mention what happens next.

Anyway, I'm flying to Italy in a few hours so have to go. Ciao!

Saturday, 18 December 2021 – Castello Steiner, Northern Italy

If I had a dollar from every person who has asked for my opinion on what happened between Lewis and Max in Abu Dhabi over the past six days, I'd be able to poach Adrian Newey! Not that I would. He's far too exciting for me. After the race I had a few days visiting my mother and every person I saw in the town wanted to know what I thought. 'Why are you asking me?' I said. 'I was too busy concentrating on a Russian who didn't finish the race and a German who was in fourteenth position.'

What do I think, then? Well, it was certainly very confusing. I remember sitting on the pit wall listening to the orders from the race director and thinking, *What the hell is happening here?* At the time it didn't stack up to me, but at the same time I didn't know all the facts. It was very entertaining, though. Poor Toto almost had a foking heart attack!

Look, at the end of the day both teams have won a world championship and good for them. Red Bull won the Drivers' and Mercedes the Constructors'. I'd take either of those. Mercedes didn't protest, so off we go. We move on.

The last few days have done me the world of good. I'm a glass-is-half-full kind of idiot so as soon as the immediate aftermath of the season lifted I started getting excited about the new one. Most people assume that after a season has finished all you want to do is relax. Bullshit! The only time I relax is when my car and drivers are performing well, which means I haven't been able to do it for over three years!

All I can think about now is the new car and all the early signs have left me being cautiously optimistic. I obviously don't know what the other teams are doing yet, which makes me nervous, but I was down in Maranello yesterday to get a feel of how things are going, and things are looking good. It's not been easy for those guys either, being stuck in an office in Italy while their team are doing shit. Because of all the Covid restrictions I haven't been able to go there as much as I usually would. Then again, if I'm there talking in their ears and making jokes, they're not working and I don't want to be a distraction. Especially not now. Fortunately, our teams in Maranello and Banbury in the UK have managed to keep their heads up and are as focused and committed as I am on making us competitive again.

I'm going to be spending Christmas at our home in Italy, which means no travelling for three weeks. That doesn't happen very often. I'll certainly be working, though, at least until the 23rd. The week between Christmas and New Year is one of the only ones of the year where hardly anything happens work-wise, so during that week I'll be forced to chill out. I have tried starting new projects during that week in the past but it's useless. Everybody's like, 'Yeah, I'll get back to you tomorrow,' but they never do.

One of the other big differences between last season and next season, which should be an advantage, is that next season we'll have two drivers who are no longer rookies. We've had to be quite careful not to be too hard on them, especially with the car we've had. Next year should be their big opportunity, though, and so we'll see what they can do. In 2021, apart from a couple of exceptions, all they've really been able to fight for is nineteenth or twentieth position. Next year I think it will be different, which will increase the team's demands on them significantly. At the end of the day, though, they are well-paid employees who need to deliver. No mercy!

Monday, 20 December 2021 – Castello Steiner, Northern Italy

Why the fok would somebody ask for an interview about Formula 1 five days before Christmas? Do these people have nothing better to do? Don't they have lives? 'You better not be asking questions about the season that's just finished.' Fortunately he wanted to talk about how Haas was created. It's actually quite a good story so I may as well tell it to you now.

You remember that period in F1 when the big manufacturers like BMW, Honda and Toyota started leaving the sport? That created a fear that there might not be enough teams so a discussion opened up about customer cars and third cars. That all came to nothing, which is when they started opening up the licences. One of the applications came from a team called USF1, which was based close to where Haas is now in North Carolina. The application was submitted in June 2009 and they were due to start competing the following year.

The team was set up by the engineer Ken Anderson and the

journalist and team manager Peter Windsor. At some point after their application had been accepted, Peter contacted me about some work that he wanted my composites company to do on behalf of the team. That's how I got to know them. Soon after that I got a call from Bernie Ecclestone. He was concerned that USF1 wouldn't be ready in time for the 2010 season and wanted to know what I thought. At first I couldn't really tell him much but after a few months it became clear that the team wouldn't be ready to start testing in January.

'No chance, Bernie,' I said. 'They haven't got a foking clue!'

Bernie went public with his concerns in December 2009 and in February 2010 Charlie Whiting visited the team's headquarters for an inspection. A few days later he confirmed that, in his opinion, they wouldn't be capable of competing and so that was that.

Anyway, one of the main investors in the team, the YouTube founder Chad Hurley, called me up after the team collapsed and asked me if I thought there was any way of getting an American team on the grid for 2011. My initial reaction was to run a mile as USF1 had been such a fiasco. Then I had an idea. Although nowhere near ready, the doomed HRT team had a car at a pretty advanced stage so I called Chad up and suggested he get in touch with the company who were building it and try and buy it.

'But I don't know them,' said Chad.

'I know,' I said. 'But I do.'

You see? Totally indispensable. How the world managed without me before I was born I have no foking idea.

'Let me have a word with the owner,' I said. 'He's a good friend of mine.'

Chad ended up flying me out to Europe to visit the company in person and see if it was doable. The meetings went OK but before I reported back to him I wanted to get the opinion of somebody who was inside the sport. The best person for that job was my old friend and fellow Italian Stefano Domenicali, who was then in charge of the Ferrari F1 Team. He invited me down to Maranello for lunch and gave me his opinion.

'You won't get it through, Guenther,' he said. 'The entire project is a complete mess. You have a good reputation in the sport so don't ruin it. Just leave it alone.'

I did have a couple of meetings with Bernie after that about resurrecting the project but the following week I called Chad up and told him to call it a day.

'It's too foking messy,' I said. 'It's up to you, Chad, but if I were you I'd walk away.'

And that was that.

A few weeks passed and, although the original project was dead in the water, the idea of an American F1 team was still a good one. Not just for whoever managed to do it successfully, but for the sport in general. It was time to call Stefano again.

'Would Ferrari consider making a customer car for a new team?' I said. 'And, if they would, would they sell it to me if I could find an investor?'

'No problem, Guenther,' said Stefano. 'You bring me the right people and I will sell you the car.'

I then had to put a business plan together. There were no lawyers and no fancy tricks, though. Just a simple PowerPoint presentation and a tall, ugly Italian with a big foking mouth.

'OK,' I said to my wife one day. 'Now all I need is a billionaire.'

A couple of weeks later I bumped into Joe Custer, who I knew from my Nascar days. He ran the Stewart-Haas Racing team and because we hadn't seen each other for a few years we chatted for over an hour. When I got home that evening I realized that I'd missed an opportunity. The owner of his team, Gene Haas, was just the kind of person I wanted to meet, so I called Joe immediately and asked what he thought.

'Do you think Mr Haas might be interested?' I said, after telling him the idea.

'Let's meet for a coffee,' said Joe. 'You can do the presentation to me and if I think Gene might be interested I'll pass it on to him.'

A couple of days later me and Joe met at a Starbucks in Mooresville, which is where a lot of Nascar teams are based, and I did the presentation.

'That's pretty interesting,' said Joe. 'OK, let me pass it to Gene. I'll get back to you when he's seen it.'

This wasn't a full-time project for me, you know. This was just a hobby. What made the presentation credible, though, was that I had Ferrari behind me. Not a bad starting point.

Exactly a month later, Joe Custer called me up and told me that Gene Haas was in town for the Nascar race in Charlotte.

'Gene wants to meet you,' said Joe. 'Let's all go out for dinner.'

This is where the project started to come alive, as Gene's interest turned it from an idea into something that could actually happen. *What if he actually says yes*, I remember thinking to myself. *Jeezoz Christ!*

At the time of it taking place I had no idea how the dinner went. Gene hardly said a word, which I now know to be normal, and I said about a million words, which has always been normal. It's the

only thing I've ever been good at and during the whole damn night
I don't think I came up for air once. Gene didn't fall asleep or get up
and walk out, though, which was a good sign.

A few weeks passed and I didn't hear a word. *I must have messed it
up*, I thought. *Oh well, back to the drawing board.* Then, completely
out of the blue about two weeks later, Gene called me one day and
asked for some more information. He still didn't say whether or not
he was interested, though, and gave me no feedback on the presentation. It was a couple of questions and was over and done within
five minutes.

Over the next few months, Gene started calling more and more,
and later on we started meeting up at his office whenever he was in
town. This went on for over a year, and although he never gave much
away (Gene Haas would make a great poker player!) the fact that he
was asking so many questions led me to believe that he was at least
interested in the project. After another six months the point had
arrived where Gene and I had to either take a shit or get off the pot.
Stefano's offer wouldn't be on the table for ever so we had to act.

'OK,' said Gene, eventually and very quietly. 'Let's do it.'

'By the way,' he said. 'How are you going to get a licence,
Guenther?'

Good question!

'Don't you worry about that,' I said. 'I'll get you a foking licence,
Gene.'

Did I know for sure that I could get one? Of course I didn't. I
was flying by the seat of my Italian shorts, as usual.

I might have been on good terms with people like Bernie and
Charlie Whiting, but that meant nothing when it came to getting a
licence.

'Get yourself a lawyer,' they said. 'You're going to need one when it comes to the licence application.'

'Why the hell would I want a lawyer?' I replied. 'I have a Guenther.'

In hindsight I should maybe have taken their advice, but I thought to myself, *No, fok the lawyers!* I didn't want to start asking Gene for money just yet and I was confident that I could do it without one. Me, a control freak? Of course I am! I'm the best there is.

In all seriousness, what I had in my favour when this all started were a lot of contacts in Formula 1 (I'll tell you how I got into F1 later). Also, unlike these days, some people liked me and I had quite a good reputation. The first person I called was my old boss, Niki Lauda. If anybody could persuade Bernie and the FIA to grant us a licence, he could. Niki was also big on the idea of there being an American F1 team, which kind of helped.

A few days after speaking to Niki, I went to Europe on business and one night while I was fast asleep in my hotel room my phone started ringing. 'Who the fok is that?' I said, looking at the screen. I didn't recognize the number but answered it anyway.

'Yes, who is it?'

'Guenther, it's Niki. I'm in India with Bernie. He's here with me now and you're on loudspeaker. He's got some questions about your licence application.'

'Really?' I said, trying to get out of bed and almost falling on my ass in the process. 'OK, yes, Bernie. Fire away.'

I had to explain at two o'clock in the morning what our business plan was to Bernie Ecclestone. Talk about a foking wake-up call! It must have lasted about an hour and a half. Thanks to this, though, Bernie warmed to the idea, which was the best possible start. After that I had a similar conversation with Charlie Whiting, who was a good friend of mine, and then after that Jean Todt, who, at the time, was the president of the FIA. I didn't know Jean very well, but with the help of people like Stefano, who put in a good word for me, he too warmed to the idea.

So, what about the business plan? Well, the original plan had been to start off with parts from Ferrari that were a year old.

'But how can we be competitive?' asked Gene. 'One-year-old gearboxes, one-year-old engines, one-year-old suspension. That's no good. It needs to be like for like.'

In the end we went to see Stefano and Mattia Binotto, who was then head of the engine department at Ferrari, and we asked them, straight out, if we could have like for like.

'Well, there's nothing in the rules that says you can't,' said Stefano. 'Sure, why not?'

For them it was less of a headache as Ferrari wouldn't have to manufacture so many parts. It made perfect sense. That, then, became our business plan. In fact, I think it was Stefano who really fleshed out the concept, which eventually became known as a satellite team.

These days you couldn't get away with it, but at the time the number of parts a customer team could buy from a manufacturing team wasn't defined in the FIA rules, and for the simple reason that nobody had ever thought about it. At the time everything was still open for discussion and in order for us not to become another Caterham or HRT we had to think differently. Either a team runs out of money, as they both did, or an investor runs out of enthusiasm. Gene and I wanted to be competitive and with the money we had at our disposal, which was obviously Gene's money and was probably on a similar scale to the teams I've just mentioned, the satellite team concept was the only solution. Everybody benefited, though: Formula 1 got a new team with a less risky future than the ones that had gone to shit, we got a competitive car and support from a manufacturer (and a pretty good one at that), and Ferrari got a new customer that, unlike some, could pay their foking bills on time. What's not to like? A lot of people in the sport complained later on, but who cares? You can't please everyone all of the time. All we did, with the help of Stefano and Ferrari, was come up with a good idea that nobody else had thought of before.

Despite all the initial positivity, Gene and I getting a licence was not a forgone conclusion. Hell, no. As I said, a lot of teams had dropped out of the sport recently so we had to prove to the FIA that Gene and I were serious and that we would not just fall by the wayside after a few months. The USF1 debacle had made them nervous and we could understand why.

What I had to do first was submit to the FIA what's called a call for interest. I don't know if you still have to do this, but it's basically just a rundown of who is behind your application and allows you to go to the next step, which is to submit the application itself. The fee you had to pay to submit the application was €150,000. It was also non-refundable, so if your application wasn't successful, tough shit. This obviously took out the bullshitters. But if your application was successful, it gave you the right to present your ideas to the panel of the FIA.

I must admit that my ass did squeak a bit while we waited for confirmation that the application had been successful. It was all good in the end, though, and so me, Gene and Joe Custer were invited over to Geneva to do a full presentation to the FIA. This is where a lawyer should probably have been consulted but it was too late by then. The presentation lasted over two hours and was basically just a set of bullet points that I talked through one by one. And there was no script. I knew the subject inside out so I freestyled. A lot of people thought I was crazy but remembering a pre-prepared script would have demanded too much of my concentration. I'm a world-champion talker, remember, plain and simple. So let me foking talk!

After the presentation had finished and the panel had all disappeared, Gene asked me how I thought we'd done. 'Do you still think we'll get the licence?' he said. 'I honestly don't know, Gene,' I replied. 'That's up to the panel. We've done everything we possibly can, though.'

Just then I received a text message from a member of the FIA panel. I'd rather not say who sent it but the message read, 'Foking hell, Guenther. Nobody can spout shit like you can! If you can't get a licence after that performance, nobody can!'

'Actually, Gene,' I said, 'let's just say that I'm quietly confident.'

Thursday, 30 December 2021 – Castello Steiner, Northern Italy

Happy Christmas!

Apparently you can buy a T-shirt online with my face on it and the words 'We look like a bunch of wankers' printed underneath. Stuart Morrison, who is our director of communications, says it's something I said on the Netflix show *Drive to Survive*.

'Don't you remember, Guenther?' he asked me.

'I can't even remember what I had for breakfast, Stuart, let alone what I said on a TV show I've never seen before. Anyway, why am I not getting any money for these T-shirts?'

Actually, that's a good question.

I've never watched a single episode of *Drive to Survive* and I probably never will. It's not because I'm against it or anything. If I was, I wouldn't appear on it. I think it's done an incredible job for Formula 1, and especially in America. My fear is that if I watch the show I won't like certain aspects of how I behave and will try and change how I do things. I know I'm not everyone's cup of tea but I'm actually OK with who I am. If you don't like it, tough shit.

Some people still believe that *Drive to Survive* is staged – or that some of it is staged – and let me assure you it isn't. You can't rehearse the shit I come out with. It's impossible! I know it's an overused saying, but what you see is what you get on there. At least with me. Some people seem to like it, which is great (they'll probably need therapy one day), and some people don't, which is fine. Seriously, I do not give a shit either way. I just go to work as normal and then sometimes a camera crew appears and occasionally they ask me questions and get in the foking way. I always try and answer those questions as honestly as possible and then I go away and carry on working. It's a simple format for me.

The question I get asked the most by the people who watch *Drive to Survive* is do I swear as much in real life. Believe it or not, I do have a filter, so if I'm in a room with some children or something, I'll try and calm it down. You know, just a couple of shits here and there. Nothing serious. In my work environment, though, it's different. I learned to speak English in a rally team and in rally it's foking obligatory to use bad language. You don't choose to do it.

Some of you might be aware of this but in 2019, I think it was, an F1 fan called Ana Colina bought me a swear box. We were in Baku and she just turned up with it one day. I forget how much I put in there in 2019 and 2020 but in 2021 it got hardly anything. I only swear when I'm excited about something, and because we weren't competitive that year I didn't have many reasons to get excited. If she'd bought me a swear box at the start of the 2018 season I'd have been able to take the whole team on a foking cruise at the end of it.

Christmas has been very quiet. Happily quiet as far as my family are concerned. Me, not so much. I don't mind a bit of peace and quiet occasionally but this is crazy. I call it 'the ceasefire' and it happens at the same time every year in F1. I'm used to getting between a hundred and a hundred and fifty emails a day and this week it's gone down to about fifteen. Somebody let me know they're alive, at least!

The only big news from Haas over the Christmas period is that our new car, the VF-22, passed one of its crash tests on 23 December. If you fail a crash test it's actually nothing too dramatic. The problem is psychological, as the people who are responsible, which in our case is our technical director, Simone Resta, and his team, are all trying to progress and look forward. If the car fails a crash test, or any test for that matter, it forces them to look back. Apart from getting the job done, there is nothing good coming out of having to do something a second time. The only things you're left with are less money and less time.

Simone joined us from Ferrari last January, where he'd been

head of chassis engineering. He's the first technical director Haas have ever had and the VF-22 project started almost in parallel with his arrival.

The test wasn't on the complete monocoque, by the way, or even half of it. It was actually just the sides. Testing it determines where the monocoque needs to be strengthened and reinforced before the final crash test takes place. Because of the nature of Romain Grosjean's accident in Bahrain in 2020, the tests have become more stringent and the FIA have increased the impact forces that the monocoques need to withstand from 20 kilonewtons (newtons are the units of force) to 30 kilonewtons. That's quite a big step.

There are literally thousands of boxes that need ticking during the process of building a new F1 car and this one gave the entire team a nice little pre-Christmas confidence boost. But especially the technical team. All they have to do now is build the foking thing!

The only other piece of news from Haas over Christmas was confirmation that Mick Schumacher will become a Ferrari reserve driver for 2022. I think it's for eleven races, with Antonio Giovinazzi covering the rest. This obviously wasn't a surprise for us. It's always on the cards when you hire a development driver, and Mattia and I had talked about it a few weeks ago. A similar arrangement was put into practice last season when Mercedes brought in George to replace Lewis, when Lewis had Covid. This shows that Ferrari value Mick, which is good. Since joining us we've been feeding them information about his progress and they're happy with how things are going. And if Mick is ever called up to replace Charles or Carlos, we have Pietro Fittipaldi to take his place. We confirmed

Pietro just before Abu Dhabi and for him this is great news as it gives him more chance of a drive. It's all good.

I'm still not sure what we're going to do for New Year yet. Whatever it is, though, it won't be too raucous. I'm too old to be getting drunk and losing a day at New Year. At my age you might not have many left! On New Year's Day we'll probably go for a walk in the mountains, but apart from that I'm going to be resting in preparation for next year. And, more importantly, next season.

OK, I'll see you on the other side.

Thursday, 13 January 2022 – Steiner Ranch, North Carolina, USA

2 p.m.

Howdy! Welcome to the Wild West.

We flew back from Italy to North Carolina on 4 January and, apart from a day trip to Los Angeles, which I'll come on to in a second, I won't be going anywhere now until the beginning of next month. It's my wife and daughter who I feel sorry for. Actually, most people who know me feel sorry for my wife and daughter! Luckily for them, our headquarters are just down the road in a town called Kannapolis. I have to be careful, though. With Covid going through the roof, it's best that I work from home as much as possible.

On Monday of this week I left the house at 5 a.m. and flew from Charlotte to Los Angeles for some meetings with Gene. I then flew back to Charlotte at 11.45 p.m. That's a 5,000-mile round trip.

Because of the time change I landed in Charlotte at 7.15 a.m. and my head was all over the foking place. I managed to work for a few hours but by the afternoon I was seeing double. That's quite normal for me, though. Travelling, not seeing double.

So, what's the news from Haas? Things are still pretty quiet at the moment but the mechanics should start travelling to Italy some-time next week. Only a handful of them but that's all we need at the moment. I've just got off the phone with Ayao Komatsu, our direc-tor of engineering. He's currently putting a simulator programme together that we'll use over in Italy on Ferrari's old simulator (they've just built a new one). The FIA have made it clear, however, that the programme we use has to be a separate entity. Otherwise we could share data. They're obviously protecting us, as I too do not want Ferrari sharing our foking secrets!

The first session on the simulator should take place with a test driver in early February. The reason we don't use the race drivers at this stage is because if there are any glitches or problems with the programme it might confuse them. After the initial sessions we'll take a view about when to bring in the race drivers but hopefully it will be soon after that.

Managing your driver's expectations is crucial at this stage. Especially young and inexperienced drivers like Mick and Nikita. As you'd expect, they're desperate to get on a simulator and it would be very easy for us just to say, 'Sure, don't you worry. We'll have everything ready for this date.' If you then can't deliver you make the driver anxious and put them on a downer about the team, which isn't healthy. It's best to tell them that you're doing every-thing you can to make it happen and then book them in only when you know that it's good to go. It might sound like quite a simple thing to do but every driver will have an entourage of managers and assistants lobbying on their behalf and if you cave in and just tell them what they want to hear, you'll be getting yourself into troble.

I won't tell the drivers this but me and Ayao are hopeful that we'll have something for Mick and Nikita to test before we go to Barcelona. How good it will be I don't know, but Ayao and the guys are doing everything they can.

Apart from that the only thing I really have to report is a couple of problems with the chassis build. It's nothing serious, though. If we didn't have any problems at all at this stage I'd be worried. It means we're progressing. The people who are probably working hardest at the moment are the purchasing department. At this time of year it's difficult to find suppliers. That can obviously be a pain in the ass at times but what it signifies to me is that the industry is busy and has plenty of work. I like busy.

The new FIA president, Mohammed Ben Sulayem, will be having meetings with all the teams soon, including us. He wants to find out how everyone is doing and what everyone thinks. A kind of state of the union thing. I've known Mohammed Ben Sulayem since my rally days, so quite a long time. He's a good guy. All the FIA presidents will have their own agenda about where they want to make a difference and I know that Mohammed is very interested in the motorsport side of things. He's a fourteen-time FIA Middle East Rally champion so it's no surprise really. Jean Todt's forte was growing the sport and he was exactly what the FIA needed when he became president. Everyone has their strengths, you know. Anyway, I'll have to mention my rally days in here somewhere along the way. Let me have a think about it.

Friday, 21 January 2022 – Steiner Ranch, North Carolina, USA

I've just spoken to Simone over in Italy and the chassis should be in the build area at Dallara on Monday next week. The workshop, the IT system and the stores are all set up now so we're ready to go. Dallara is the race car manufacturer that has been developing the car with us. We've been working with Dallara since 2016 and they're a great bunch of people. The main areas they work with us on are aerodynamics, vehicle dynamics, and the design and the structural calculation of the car. They also help us to incorporate all the various components with the engine and transmission. It's a good relationship and I've known the owner of the company, Gian Paolo Dallara, for decades. The problems will come thick and fast as soon as they start the build, but that's normal. This isn't my favourite time of the year but it's one of them. It's when all of our dreams start to become a physical reality.

Apart from what's going on in Italy, the thing that's been taking up most of my time this week has been budget stuff. Sorry, what I should have said is, 'The thing that's been sending me to sleep this week!' It's the proverbial necessary evil, you know, and all week I've been back and forth with the board getting things signed off. It's all loose end stuff really, so at least I know it's coming to an end. The big push with regards to activity will start the week after next. In the meantime, though, once the car build begins I'll start getting a load of moaning telephone calls and emails asking why we don't have this, that or the other. Half of my life is spent dealing with this kind of shit but do you know what, I love it. Why? Because if people weren't moaning nothing would be happening and I'd have nothing to do. It's that foking simple.

Thursday, 27 January 2022 – Steiner Ranch, North Carolina, USA

2 p.m.

Today I'm quite excited because tomorrow I am flying down to Florida for the 24 Hours of Daytona. It starts Saturday afternoon and finishes Sunday. I'm only going for the day (it's just a two-hour flight) and to be perfectly honest with you the main reason I'm flying down there is to catch up with some friends. Kevin Magnussen is going to be driving, which is good. I haven't seen him since he left Haas. Hopefully he'll still be talking to me. I still have to message Romain to see if he's going to be there. He won't be driving but he lives in Miami so you never know. It's only a four-hour drive away and in America that's fok all. The three of us could have a little reunion! I actually miss those guys, though. We went through a hell of a lot together.

Romain has been doing very well in IndyCar since he left. I must admit it's been a little bit of a surprise how fast he's been because it's not easy. You really need to drive those things. I thought he'd struggle in the beginning but he's proved me wrong. Then again, the fact that they have no power steering and are difficult to drive probably appeals to Romain. I always said that, on a good day, Romain could have been world champion. He just could never get the consistency right. Also, when he had a bad day, he *really* had a bad day. You could probably say that about a lot of drivers, in that they can beat anyone on their day but their day doesn't come around often enough. Romain would also try too hard sometimes. At the end of the day, a driver cannot make up the deficiency of a car. Or at least, not to any great degree. He used to try and do things that the car wasn't capable of, which is when things used to go wrong.

Kevin was very young and very immature when he came into Formula 1 and in my opinion he would have benefited from having one more year working with a mentor. Also, I think he was in the wrong team at the beginning. The Magnussens are old-fashioned racers and they need a bit of freedom. McLaren was never going to give that to Kevin and I think it held him back. Later on he got a bad reputation and then he was out for a year. When that happens, and especially at such a young age, you're bound to lose a bit of your confidence. That's why he did so well with us, I think. We gave him the freedom he needed and we supported him. Look, as long as you drive a good race I'll give you all the freedom you want. As long as you do exactly as I say and don't foking crash!

I really do hope that the three of us get to have a chat together. If we do, we'll spend twenty minutes reminiscing and then Kevin and Romain will spend an hour taking the piss out of me. That, I can guarantee you. They're masters of it.

People often ask me what I like to do in my spare time and when I tell them I like to jump on planes and go and watch car races they can't believe it.

'You go to a car race? What kind of drugs are you taking?'

'But I don't know anything else,' I say to them. 'What am I supposed to do?'

I do like other things. Of course I do. But there's nothing in my life apart from motorsport that would make me get on a plane or buy a ticket. I'm not the only one. Motorsport takes over your life and I've never known anything else. Since 1986 that's all I've been doing and it's where all my friends are. Just this morning I spoke with my old boss, Malcolm Wilson. His team, M-Sport, which I used to run, had won the opening round of the 2022 World Rally Championship in Monte Carlo the other day and we had a good old chat about it. Malcolm is one of four people – Malcolm, Niki Lauda, Carlos Sainz Sr and Gene Haas – who have had the biggest influence on me in motorsport and I owe him a hell of a lot. It can get a bit incestuous in motorsport sometimes, but as long as we're not breeding with each other, who foking cares?

I once had to spend two hundred days a year with Carlos Sainz.

OK, what's been happening in Haas world? Well, the build has begun as promised, which is good. The chassis was a day late coming back from England, where it had been painted, but there you go. There was a mini panic about that at Dallara. And about some parts that are running late. Because of Covid, everything is running late this year. It's just one of those things and I'm not worried. The people in Italy are, but that's part of what they all do for a living. In fact, I might put that on their business cards next time.

'Simone Resta – Technical Director at Haas F1 & World Champion Worrier.'

We'll get there in the end.

This morning I reminded my wife and daughter that this is my last week at home before everything starts again.

'Next week the circus begins,' I said.

They didn't say very much but I could tell what they were thinking. 'Ten months of peace and quiet!'

Most people cannot get their heads around the arrangement that I and everyone else who lives this kind of mad nomadic life has with their families. I met my wife almost thirty years ago so I was already on the road in motorsport at that point. Me being away from home is perfectly normal, then, whereas me being at home is actually abnormal. She and our daughter Greta have never known any different. I do sometimes wonder what it would be like if it was turned on its head and I was at home all the time. I don't do it for very long, though. After about thirty seconds I start sweating and my head begins to hurt. It's terrifying!

PRE-SEASON

Saturday, 5 February 2022 – Maranello, Italy

Daytona was fun. I usually go there every year but because of Covid I haven't been for the last two. It was great to be back. I only stayed for six hours but would happily have stayed for two days. And I did some serious talking. I really am the King, you know. In fact, fok the 24 Hours or Daytona. How about the 24 Hours of Guenther? That would be a real challenge!

As much as I am addicted to motorsport, I must admit that freeing myself from the world of Formula 1 for a few hours was fantastic. It's something that I don't get to do very often and it always happens just at the right time. That's probably why I talk so much when I'm there. You see, if I stop talking and stop asking people questions, they will start asking me questions about Formula 1, and that would spoil everything. I have to keep talking! Also, the atmosphere at Daytona is completely different to Formula 1. Formula 1 can be quite serious and businesslike sometimes, whereas Daytona is completely laid-back. As soon as I arrive there I can feel the weight on my shoulders disappearing. It's amazing.

Funnily enough, the first person I ran into when I arrived in the paddock this time was Kevin. He was driving a golf cart, which looked to be much too powerful for him. 'What are you doing?' I said. 'You're all over the foking place!' I think we spoke for about twenty-five minutes and it was great to catch up. I obviously can't speak for him but from my point of view I definitely think we're friends again. One of the first things Kevin said to me was that he doesn't think he could have driven for us last season, given that we weren't competitive. 'Which is exactly what I said to you when I let you go,' I said to him.

By the time I informed Kevin and Romain that we wouldn't be keeping them on in 2021, Gene and I had already made the decision to run the 2020 car in 2021. I tried to impress on them both the effect that racing an old and uncompetitive car would have, not only on their careers, but on their state of mind. I don't think it sank in at the time but once they saw the stark reality as the season unfolded they realized that, actually, Guenther might have done them a favour. Kevin's got a good job now with the Chip Ganassi Racing team in sports cars in America, and in Europe with the Peugeot WEC team, alongside Paul di Resta and Jean-Éric Vergne. In fact, the race at Daytona was his debut with Chip Ganassi and if it hadn't been for a puncture while running P2, he would probably have started off with a win. He's also been doing a lot of F1 work back in Denmark, as well as some TV commercials. I'm really happy for him. He's still quite young so who knows what might happen for him in the future.

I arrived in Italy on Tuesday the 1st. Our office in Maranello, which is where I work from when I'm over here and where I am now, is based within the Ferrari headquarters but is a completely separate facility. It has to be. As I just said, I can't have those bastards stealing all of our secrets! In 2020, we decided to review our operation, including what we were doing on the technical and development side. Gene always said that he would review his decision whether or not to stay in the sport every five years, and as we started the team in 2015, 2020 was the time. Fortunately, he decided not to send me on to the streets and when that decision was made I started to review what we were doing.

When the budget cap came in, some people at Ferrari had to be let go. It must have been the same for a lot of teams, which was sad, but in this case it presented an opportunity. At the time we were still based at Dallara and so, after meeting with Mattia, who has been at Ferrari since before he was born, it was suggested that we move our operation to Maranello and take on some of the people who had been laid off. Ferrari had built a new facility before the budget cap, which would now be standing empty, and because it was shielded from their own F1 operation it was perfect for us. We still have some people working at Dallara but nowhere near as many. I must say, Maranello isn't a bad place to work. I have more friends here than I can remember and the food is foking excellent.

Our relationship with Ferrari has always been very positive and it continues to evolve. What prevents it from getting stale, I think, is my relationship with Mattia. We've been friends for a very long time, and when issues appear, which they inevitably do from time to time, we always know how to sort them out. Or should I say, he does. I'm aggressive and stubborn, whereas he is conciliatory and almost horizontal. It just works.

The new simulator programme is coming along well and should be ready for Mick and Nikita in the next week or so. Last week our director of engineering, Ayao, came over and did some sessions with a test driver, and apparently they went really well. There's still some development needed and they are sorting out the few remaining bugs. Shouldn't be long now, though. I think Nikita is in first so it will be interesting to see how he gets on. Mick will hopefully get a go in before Barcelona but, if not, he'll definitely go in before Bahrain.

Yesterday was a big day for Haas as we became the first team on the grid to reveal our car and livery for the new season. It's obviously only a digital image, but it's the first glimpse the press and public have had of an F1 car that has been designed using the new technical regulations. These, if you aren't aware, include things like 18-inch tyres, a ground effect floor, which helps the cars stick to the ground, and a simplified front wing and dramatic rear wing. I won't go into the finer details of the regulations as it would put you to sleep. For the fans, though, the first launch of the year is like the first day of Christmas and everyone starts getting excited for the season ahead.

According to Simone, in terms of creating a new car, it's the most complex project he's worked on in his twenty-year career. 'I can't remember such a big change,' he said to me. The digital images look fantastic but there's still a lot of uncertainty about what to expect in Barcelona.

So why have we launched so early? Well, when Haas started out we were always the first team on the grid to launch our new car. As the smallest team, it was our opportunity to be in the spotlight for a while and get some publicity, and after a while it became expected. Unless you are one of the big teams, nobody really cares too much about your launch and so for us it was always important. Last year, that fell by the wayside as we were launching a car that had basically already been launched. What's the point of being first with something old? You just look stupid. Now that everything is new again we decided to get our crown back.

There's actually quite a lot involved in the process of launching a new F1 car. It's not just a case of creating a digital image of the car you are building and then sending it out with a press release. That's how it used to be in the early days but now things have changed. These days, everything has to be signed off by an army of different people. The board, the sponsors, the marketing people. A lot of boxes need to be ticked first and there will always be somebody who isn't happy and wants something changed or has questions. The earlier we can get this done, though, the better it is for the team, as it's one less thing to worry about.

The most challenging part of the process is releasing an image that satisfies the press and the public's desire to see something new, but does not give too much away in terms of geometry. If you don't get that right you might give something away to the competition. I think we've got that right. Or at least I hope we have.

The only other thing I need to report on at the moment is how the car build is going. They've been waiting all this week for some parts to arrive (which are late) and so next week is going to be like a pressure cooker. Now that the launch is over, at least I can concentrate on helping the guys find solutions to any last-minute problems. That's probably the thing I'm best at doing within an F1 team and I really quite enjoy it. We are running a little late, though, which is a bit of a worry. Providing those parts arrive when the suppliers now say they will, we should be OK. If not . . . Well, let's just say that some shit will be hitting a few fans and it could get messy!

I'm now in daily communication with Ayao, Simone and Stephen Mahon, who is the global programme manager at Haas and in charge of the planning and organization of the car build. The next two weeks are going to be crucial and I need to know exactly what's going on. Especially if there is an issue. Once again, my talent, if you like, is helping to prevent small issues from becoming big ones. I've been doing it a long time and I think most of the guys at Haas respect my experience and like the fact that their team principal is a hands-on kind of guy. At the end of the day, they have no foking choice in the matter, so there you go! I also need to be able to report back to Gene so, like it or not, I really do need to know everything.

My other main job now is helping the team to prepare for the race weekends. The list of things that need to be done is as long as an elephant's trunk and because of the size of our team (which is considerably smaller than the other teams), and because I am ultimately responsible for the budget, I have to be involved in almost every aspect. Some people might accuse me of being a control freak but when you've been running things from day one and are partly responsible for the team coming into existence in the first place, it's hard not to be. What prevents me from becoming a complete pain in the ass is that I always try and surround myself with good people. They are the ones who keep me sane from day to day and, as long as I know that they're on top of things and are doing a good job, I can take a step back, leave them alone, and concentrate on something else.

Last week a new marketing director started. Normally, having such a senior member of staff coming in at such a busy time of the year would be a bit of a struggle, but not this time. Before his arrival I had always been the de facto marketing director for Haas F1 and, to be honest with you, it wasn't something that I particularly enjoyed doing. I did it to the best of my ability but I was definitely a fish out of water. Marketing people have all kinds of qualifications these days and I haven't got any qualifications for anything! He's only been in the position for a week or so but already he's been making a difference. I haven't asked him if he's had any problems clearing up all the shit that I left behind yet, but I'm sure he'll let me know.

Friday, 11 February 2022 – Maranello, Italy

Aston Martin launched their new car the other day. I've spoken to our guys and there are no great surprises. There are still eight more launches to go but we won't know anything about any of them really until we get to Barcelona. Our own car is coming along really well. There are lots of issues, but they are all positive issues. Or issues that are associated with a car you are building from scratch as opposed to a car you are just re-launching. If I am not panicking about something at this time of year then I know that something is wrong, and I'm panicking a lot at the moment. It feels like the good old days and although we're being cautious there is definitely a feeling around the team that good times might be ahead. That's what I've been promising the guys for the past year but to actually feel like it might be happening at last is fantastic. Everybody's getting very excited now. They're pumped up and ready. The atmosphere that hope creates really is the lifeblood of every sport.

Developing a car is a funny business. It's addictive, challenging, frustrating, infuriating, rewarding. A permanent struggle. The further along you push it the more development you can do and the better the car should be. In rallying, the deadline for development is normally homologation date, whereas in F1 it is your first test. This means you can carry on developing right up until your car leaves the garage for the first time, which in turn allows you to get the most out of the regulations. And you have to use that time. All of it. If your car is ready a month or even a week before the first test you are doing something wrong. Every minute counts.

The budget cap has had a much deeper effect on the bigger teams than the smaller ones because we are already used to not just working with less money, but getting more out of it. Hopefully, that will give us an advantage.

One of the most common questions people ask me about my days in rally is what do I miss about it. Well, apart from the people, who I miss a lot, the thing I miss most about rally is the variety of experiences you get when you are travelling. With Formula 1 it's race track, hotel, race track, hotel, race track, hotel, race track, airport. Sometimes you might get to go for a meal at a restaurant on the Thursday or Friday night but usually you just divide your time between the race track and your hotel. I think musicians get the same thing. If you're on tour, the chances are you'll be playing in a city so it's venue, hotel, airport. You never get to see the city or country. Or not to any great degree. In rally, we used to test for a month in Africa and every night we'd stay in a different hotel in a different place. As well as seeing and learning about each country, we got to meet the people, which is another thing you don't get much of with Formula 1. Hotel receptionists and waiters are the people we most come into contact with.

The other thing I miss about the early days of my rally career is the lack of organization. It was all very free and disjointed when I started back in the 1980s, and one of the reasons for that was because we didn't have cell phones. You were given a car, you went testing, and at the end of every day you'd fax back what did and didn't happen. That was the only way you could communicate a lot of the time, so if you needed something like a new part and couldn't get a phone line, which was quite normal in Africa, that's what you'd have to do. This eventually changed and with it so did the demeanour of the drivers. These days the demeanour of a rally driver is far more in tune with an F1 driver. Primarily because they're well looked after and everything around them is organized. They're still fruitcakes, generally. It's just a requirement of the job. In Formula 1 you have to be very fit and in rally you have to be slightly mad.

The variables in rally are far greater than in F1 and because of that it's a more reactive sport. And I'm not just talking about the terrain. I'm talking about the weather and the climate, too. Look at the Monte Carlo Rally. One stage could be icy and the next completely dry. Self-sufficiency is also key to being a rally driver. In F1 you will have a team of engineers telling you what you need to know and what you need to do, whereas in rally it's just you and your co-driver. The breadth of talent required is wider, I'd say.

My only other issue this week, which I really do have to get done before Barcelona, is sorting out the last few sponsorship contracts. As opposed to having a sponsorship director in house, we tend to use agencies who do the legwork on our behalf and then make an introduction. And who do these lucky people get to meet and negotiate with if they show an interest in sponsoring Haas F1? Me. I always assume that this might put people off but I'm doing OK at the moment. Nobody has screamed yet or called the police.

If I don't get them sorted before Barcelona then it won't be the end of the world but it has to be before we go racing.

On Sunday I'm flying to the UK. We have a factory in Banbury and there's a lot going on there at the moment. Actually, there's a lot going on everywhere at the moment. While I was moaning about not having enough time the other day somebody suggested that I clone myself. 'Are you foking crazy?' I said. 'I'd just end up arguing with myself all the time. It would be a nightmare!' Multiple Guenthers? *No grazie.*

Friday, 18 February 2022 – Maranello, Italy

On Monday the 14th the F1 Commission gathered in London for its first meeting of the year. It was the first meeting attended by the newly elected FIA president, my old friend Mohammed Ben Sulayem, and there was a lot to talk about. The first item on the agenda was last season's Abu Dhabi Grand Prix. Jeezoz, not again! How many times do we have to relive this? I'm not allowed to tell you what was said in the meeting but let's just say that a full and frank discussion took place between certain members of the commission that kept me entertained. Do you know, I'd happily have bought a ticket for that. Fortunately, I didn't have to. Front row but no popcorn.

Next on the agenda was the sprint races. First we had a review of the three that took place last season and everybody was in agreement that the new format had been a success. Three more were also proposed for the new season at the Emilia Romagna Grand Prix, the Austrian Grand Prix and the Brazilian Grand Prix. There'll be a couple of changes to the points system, I think, but apart from that it's all the same. I think they're great for the sport as they offer additional value without us having to extend what we are doing by much. I'd like to see more of them, to be honest, rather than more Grands Prix.

The only other major thing we covered was what should happen when races are shortened due to weather. Following what happened in Belgium last year, which ended up being a washout, the commission have proposed some updates to the regulations such as points only being awarded when a minimum of two laps have been completed by the leader without a safety car and/or virtual safety car intervention. I'm in agreement.

So, Michael Masi will no longer be the race director, which is obviously because of what happened in Abu Dhabi. I think the decision by the FIA to keep him involved in the sport as opposed to just getting rid of him is the right one. Because of what happened there's no way he could have continued as race director. Whether you agree with what he did or not his position would have been under far too much scrutiny and every decision he made would have been questioned again and again. I haven't spoken to anybody at the FIA about this so it's just my opinion. I like Michael, though, and regardless of what happened last year the level of criticism and abuse he has had to face has been ridiculous. It's really not good.

On Tuesday I spent the day at our factory in Banbury, where I saw Nikita briefly. I always try and keep my distance from the drivers as much as I can in the off-season because during the season I see them all the time. He seems OK, though. He had a good time in the simulator last week and the programme seems to be coming along.

On Wednesday I flew back to Italy and after spending the day at

Dallara I drove down to Maranello, which is where I am now. Yesterday morning at 5 a.m. we fired up the engine for the first time. It was a few hours behind schedule but it sounded foking amazing. That's always a good feeling, you know, and everybody's very happy. How fast will it be? We won't know for another week. We're almost there, though, and tomorrow morning the car will be transported from Italy to Barcelona. As I sit here I can actually hear the new Ferrari in the background. Our office looks directly on to the test track in Fiorano but the blinds are closed so I can't see anything. It's almost as if they don't trust me.

I still haven't sorted the sponsors out completely. There are just a few discussions going on about where we are going to put logos but we're almost there. In fact, I've got a couple of calls to make about it this afternoon so I could actually have it sorted out today.

The really big news this week, not just from a Haas point of view but motorsport in general (not to mention fashion), is that a video has been released on social media by our social media team of me modelling some teamwear. Except that instead of it just being me standing with my hands in my pockets looking at the camera, somebody has put 'Careless Whisper' by George Michael over it! All of a sudden people started laughing in the office and then the next thing I know I get an email from my business partner in the States with a link saying something like, 'Look, Guenther, you're a supermodel!' The mistake I made was actually clicking on the link. 'What the hell is this?' I said. There are some things you cannot unsee. I'm traumatized! A few minutes later I get a text from my daughter. 'Looking good, Dad!' she said. 'You make a great model.'

Why do people do these things to me? I have feelings too, you know!

Anyway, next stop Barcelona. I feel really nervous, to be honest with you. Me and the team have done everything we can to prepare a car that will make us competitive again and the moment of truth is almost here.

TESTING

Tuesday, 22 February 2022 – Circuit de Barcelona–Catalunya, Barcelona, Spain

9 a.m.

I arrived at the track at 6 a.m. this morning and the first person I saw was Stuart, our director of communications, looking worried.

'Guenther, I need to speak to you,' he said.

I thought to myself, *Shit, here we go*. The test hasn't even started yet and already we're having problems.

He told me that Russia might be on the verge of invading Ukraine. To any other team that wouldn't be a problem, as such, but to us it could be disastrous. Not only do we have a Russian driver but our title sponsor is also Russian. I don't even want to think about it at the moment.

About an hour later I saw Nikita. Everybody's trying to act as normally as they can around him but as elephants in the room go this one is the size of a foking mountain! At least preparations in the garage are going OK, which is something. The main problem we're expecting at the moment is the car porpoising on the straight, which results in bouncing, and is to do with how the new cars are creating downforce. It's ground effect, basically, so as the car is sucked towards the track, it creates more downforce. As the car reaches its top speed, however, the ride height of the car decreases, which can cause the airflow to be lost under the car, causing the car to lift. As the car then creates downforce again the process starts anew. Anyway, that's the technical team's problem at the moment. Let's hope they can sort it.

The rest of the day was spent catching up with the guys and doing interviews. Everybody wants to know about the Russia situation but what can I tell them? I'm not Vladimir Putin! I kept saying to the reporters, 'Do you want to talk to me about the car or the team? I mean, that's why we're here, isn't it?' They got the message.

It's good to have everybody back together again. I just hope the car is as good as we think it is.

Wednesday, 23 February 2022 – Circuit de Barcelona–Catalunya, Barcelona, Spain

2 p.m.

I slept like shit. My brain is like a hamster's wheel and there are too many damn distractions at the moment. It was still dark when we set off to the track so I tried to sleep a bit more. I had some breakfast when I arrived and then went straight to the garage. At least the weather's OK. You have to look on the bright side.

Nikita was the first driver to go out but after twenty laps we discovered a cooling leak and lost valuable track time trying to fix it.

This is just foking typical of my luck. He returned to the circuit before lunch but only for a few laps.

As first mornings go that was a shit one and the afternoon didn't go much better. Mick spent most of the session sitting on his ass because of a damaged floor and he only managed twenty-three laps. Only Bottas and Kubica at Alfa Romeo did fewer laps than our two drivers. Here we go!

You know, earlier today a friend of mine said that if there was a turd in the middle of any road in the entire world I would end up standing in it.

'Guenther,' he said to me. 'I like you but you're a walking disaster zone!'

This is a good friend of mine so imagine what the guys who don't like me say! He was only kidding around but for the last three years everything I've touched has turned to shit! I said to my friend, 'It's not all my fault, you asshole. I just can't get a break!'

After the test, I had a meeting with Stuart and then a very long telephone call with Gene. He's worried about the whole Russia situation and has set up a meeting with the board of Haas Automation tomorrow after the test. I've also had a couple of sponsors on the phone about this, wanting to know what we're going to do about it. Everybody wants answers but I can't provide them. Stuart has suggested that I lie low for a while and don't speak to the press, which is a good idea. It'll be nice to have a break.

Early night. I'm exhausted!

Thursday, 24 February 2022 – Team Hotel, Barcelona, Spain

10 p.m.

Jeezoz Christ. What a day! I've had to make notes so I don't forget anything.

I woke up to the news that Russia has now invaded Ukraine. Oh great! I obviously feel very sorry for the people who are directly affected by this but I can only look after my own ship, you know? When it comes to motorsport, all eyes are on us at the moment. I didn't even turn my phone on until after I got to the track this morning as I knew it would be ringing off the hook. When I eventually turned it on there were over a hundred texts and about seventy voice messages.

The first person I saw when I arrived was one of our engineers, who immediately made me laugh.

'Guenther,' he said. 'Only Haas could have a Russian driver and a Russian sponsor at the start of a Russian war that makes everybody else in the world hate Russia!'

I almost wet my trousers. He's right, though. We have some kind of curse on us, I think. Perhaps Kevin and Romain are to blame? I thought I'd made it up with Kevin but you never know. They could be sitting at home right now sticking pins into Guenther dolls!

He might have been joking but the engineer was right. With a Russian sponsor *and* a Russian driver, everybody wanted to know how we were going to respond to the invasion. It was pretty intense and we made the decision first thing to say nothing to the press until I'd spoken to the board. The first thing I had to do after that was make sure that the guys in the garage could carry on testing without any outside interference. That was easy up to a point, as they're all very focused, but I had to have a very difficult conversation with Nikita. I know that his father, Dmitry, who is the majority share-holder of our main sponsor, Uralkali, is close to Vladimir Putin and, at the end of the day, I don't want our team to be associated with someone who starts a foking war, you know? Nikita said that he wasn't interested in politics and just wanted to drive. I under-stand and appreciate what he's saying but it's a bit bigger than that. It's so difficult for everyone.

It ended up being a pretty good day on the track. Not brilliant, but better than yesterday. Mick went out early, set a fastest lap time of 1:21.949 and accumulated sixty-six laps, which is the same dis-tance as a full Spanish GP. He seemed happy. Seven laps into Nikita's run he stopped on the track with a damaged fuel pump so the session was red flagged. The guys worked their foking asses off to fix it and on his penultimate lap he set 1:21.512 and brought the total lap count for the day up to 108. Nikita will commence run-ning on Friday morning and Mick will bring the test to a close in the afternoon.

What's keeping the guys in the garage focused at the moment is that they can all see potential in the car. And the difference between this and everything else that's going on is that the potential with the car is real. It's happening. Russia invading Ukraine is also happening but there's nothing we can do about that. Everything we've had thrown at us since the invasion started has been opinion and it's been relentless. The two things that every adult has in common is an opinion and an arsehole and when the two meet, bullshit suddenly appears.

As soon as the test was over I went to my office for the board meeting. They wanted to know what I thought, as team principal, so I told them. 'Drop the Uralkali branding,' I said. 'Change the livery to white and tell the whole foking world that is what we have done.'

Every single board member agreed with me and so the meeting was over in a couple of minutes. There was no debate to be had. It needs to happen for the good of Haas and for the good of the sport. If we retained Uralkali as a sponsor and had them on our livery we'd be crucified by the media, the fans and the FIA. It would be suicide and I've got enough on my plate!

After the board meeting I called Uralkali's chairman and told him what we'd decided. I think he'd been expecting it. I could have called Nikita's father but every time we speak we end up having an argument. Although the decision was unanimous, I asked the chairman of Uralkali to tell me in his own time how they saw things. I still haven't heard back from him. The whole thing is what people these days call a very 'fluid situation', which basically means that nobody knows shit about what's going on. I hear it all the time. It's trendy these days.

'I don't know how to do my job any more.'

'Hey, don't worry, man, you're just in a very fluid situation!'

What a load of bullshit.

Anyway, at 6 p.m., Stuart released the following statement to the press:

Haas F1 Team will present its VF-22 in a plain white livery, minus Uralkali branding, for the third and final day of track running at Circuit de Barcelona on Friday 25 February. Nikita

Mazepin will drive as planned in the morning session with Mick Schumacher taking over in the afternoon. No further comment will be made at this time regarding team partner agreements.

Too foking right there won't be! Just leave me alone.

That was as much as we could do today but according to Stuart it did the trick, in that it helped to pacify the press, the fans and the FIA. They don't know that we haven't actually terminated our relationship with Uralkali yet but this will keep them off our backs until we can go public. As Gene said to me last night, 'It's all about damage limitation at the moment.'

What I intend to do now is get through tomorrow, fly home and then lie low for the weekend, see my family and keep my big mouth shut. Yesterday and today I didn't say one word to the media and the rumour mill has been in overdrive.

'Guenther's been sacked,' they said. 'Haas have gone bust!'

More bullshit.

I also had Netflix knocking at the door all day. According to Stuart, the new series of *Drive to Survive* began a few days ago, which means half the grid will be moaning about it, which is what always happens, and the other half won't give a shit. It's just a television programme and it's good for the sport.

Given what has happened over the last few days, I have high hopes for the new series of *Drive to Survive*, as it will be something other than Russia and Nikita that people will want to talk to me about. At least, I hope that will be the case. I'm OK talking about most things but politics and shit like this, less so.

The rumours about Haas going under if Uralkali are dropped as a sponsor are the first ones I will respond to when we decide to go public, and for the simple reason that a lot of our fans and support-ers are genuinely worried about the team's future. We're in a pretty good place financially so there's nothing to worry about. Life is all about overcoming obstacles and I've already made a few telephone calls. We'll be OK. The most important thing at the moment, now that I've spoken to Uralkali, is making sure the team don't feel threatened. Apart from one of our drivers, who probably is shitting himself a bit, I think that's the case. Unlike the other team princi-pals, I've been involved in recruiting every member of staff we have on our team. They trust me and we're like a family. So what if it's the Addams Family?

As I write, the FIA are meeting to decide what to do about Rus-sian drivers and the Russian Grand Prix. I've already had members of the World Council on the phone this evening asking for my opinion. The only reason they want my opinion is to justify their own and I won't give it to them. I don't like being used. It's a deci-sion they have to make on their own. If they come out and say no Russian drivers at any Grand Prix then that will make my life a lot easier, but let's see. At the end of the day the best outcome for eve-ryone would have been Russia not invading Ukraine in the first place.

When I got back to the hotel a couple of hours ago I rang Gertie, who fortunately didn't ask me too many questions about Russia. She's been worried about me, though, so I told her that everything is fine. I had something to eat and then answered some emails and messages. On a shit scale from one to ten, today has been about a million.

Seriously, who'd have my life?

Friday, 25 February 2022 – Circuit de Barcelona-Catalunya, Barcelona, Spain

4 p.m.

As soon as I got to the track this morning, everything began turning to shit, for a change. Literally the moment I walked into the garage a mechanic started complaining that they'd found a leak in the oil system and we only managed nine laps in both sessions. What it means in real terms is that we've had one full day in a three-day test. That's just the way it goes sometimes. The main thing is we've learned a lot about the car and have gone some way to solving the porpoising. I think 80 per cent of the teams have had the same problem.

The Russia situation is obviously ongoing but the response to what we've done so far has been very positive. I wasn't sure how it would go at first but I think we're happy. The FIA have also released a statement about the Russian Grand Prix, which is no longer going ahead. No word on Russian drivers yet, though. At the end of the day there is only one Russian driver in F1 and, surprise, surprise, he works for me! That's typical. We'll see. Right now I just want to get home and see Gertie and Greta.

PRE-SEASON

Wednesday, 2 March 2022 – Steiner Ranch, North Carolina, USA

2 p.m.

Ever since the press release went out the other day, every driver with a foking super licence has been trying to kiss my ass. I've always been a super-popular guy, you know, but when people think you've got a seat in Formula 1 to give away that popularity gets a turbo put on it. I had a text about ten minutes ago from a driver, who will have to remain nameless, asking me straight out if I was going to drop Nikita. And if I do, could he have a word with me? I'm like, *What the fok? Are you crazy?* I'm getting a lot of love at the moment. It's Guenther time!

The Nikita situation is my biggest worry just now and is causing us a lot of problems. Even if we kept him on, because of the sanctions that have now been placed on Russian nationals in some countries, he wouldn't be able to take part in all the races. Then what if Mick gets called up by Ferrari one week? I could potentially end up with one reserve driver. We're going to have to make a decision in the next couple of days. If we let Nikita go, we'll have to find a replacement quickly.

Saturday, 5 March 2022 – Steiner Ranch, North Carolina, USA

8 p.m.

On Friday the 4th, Gene and I made the decision to officially sever ties with Nikita and Uralkali. It obviously wasn't a difficult decision to make but, before letting them know, we had to do a lot of work with the FIA. It had got to a point where our other sponsors were going to leave Haas if we didn't act now and had we waited any longer we could have ended up with no sponsors at all.

Later that afternoon I drove to Ashville in the west of North Carolina with my daughter for a swimming competition that she was taking part in the following day. Then, at three o'clock the next morning, while she was fast asleep, I got up for a meeting with Stuart, who was in the UK. The amount of work we had to do before announcing the decision was off the scale and we also had to write letters and send emails to Uralkali and Nikita informing them. It was a hard slog but by about 6 a.m. everything was done and ready to go, which meant I could go back to bed. The statement was deliberately short:

> Haas F1 Team has elected to terminate, with immediate effect, the title partnership of Uralkali, and the driver contract of Nikita Mazepin. As with the rest of the Formula 1 community, the team is shocked and saddened by the invasion of Ukraine and wishes for a swift and peaceful end to the conflict.

We arrived back from Ashville this afternoon and as soon as we landed I went straight into my study to start tackling the fallout from the announcement. Because I've been with my daughter my cell phone has been switched off, and when I switched it on again it went crazy for about ten minutes. It was as if it was saying, 'Where the fok have you been, Guenther?'

Stuart has fielded a lot of the emails but the people who know me personally have all been trying to talk. What can I tell them? They know the reasons why we have decided to sever ties with Uralkali and Nikita. There's nothing more to say. As the day has passed, more and more people have been asking who is going to be replacing Nikita and that is what is on my mind at the moment. In fact, I have a meeting with Gene about it first thing tomorrow morning. To be honest, I'm still not sure what we're going to do. As reserve driver, Pietro Fittipaldi would seem like the obvious choice but, despite making a couple of starts for us replacing Romain, he would basically be a rookie. What I think we need now more than anything is experience – and Gene feels the same. The final decision will land with him but at the moment there's no obvious successor.

Sunday, 6 March 2022 – Steiner Ranch, North Carolina, USA

5 p.m.

Gene Haas likes to lob out a good idea sometimes and this morning, when we had our meeting about who would replace Nikita, he surpassed himself.

'How about getting Magnussen back?' he said. 'Do you think he'd do it?'

'Jeezoz Christ, Gene,' I said to him. 'You're a foking genius, you know that? We had a good talk at Daytona a few weeks ago and my opinion is that if he can, I think he will. Let me speak to him.'

To be honest with you I had absolutely no idea whether or not Kevin would come back to Haas, but I fancied my chances of persuading him to say yes. The two things I'm good at other than talking bullshit all the time are delivering bad news and persuading people to say yes. That's my entire skillset. In fact, if I ever have to look for another job I'll put it on my résumé.

Name: Guenther Steiner

Date of birth: A foking long time ago

Skills: Bullshit, bad news, intense persuasion

Talking of having to deliver bad news, even before we'd made the decision to contact Kevin, Gene and I had decided not to offer Pietro the seat. He took the news OK but he was obviously very disappointed. Look, I don't want to disappoint anybody but I have to do what's best for the team and that's all there is to it. We need to get back to where we were and I cannot take any more risks. I don't need another rookie.

I should point out that Gene suggesting Kevin for the seat wasn't just a stab in the dark. First, we know Kevin well and he knows us. Despite the car being new, the majority of the team are established and so for a driver to be able to walk into the garage, shake a few hands and hit the ground running would be amazing. And then there's Mick, of course. Kevin already has well over a hundred Grand Prix entries to his name and Mick would only benefit from his experience. It also might help us ascertain how good Mick is because at the moment we don't know. Next, everybody at Haas likes Kevin. He's completely apolitical and, like us, he just wants to go racing. Last but not least, what better way for Gene and me to demonstrate to all the guys in the team that we are serious about getting back to the good old days than by re-hiring one of the drivers – from the good old days.

By the time I picked up the phone to call Kevin earlier, I'd convinced myself that he was the man for the job.

The response from Kevin when I called him was typically understated. He's quite a quiet and reserved character and, after I'd asked him how he'd feel about coming back to Haas and told him why we thought it would work, he said something like, 'Yes, OK then.' I thought, *Wow, I'm glad you're as happy as I am, Kevin!*

We still need to work out all the details and get him released from his contract with Chip Ganassi and Peugeot, but I'd be very surprised if they tried to stand in his way.

What a great way to end a very difficult weekend. The future starts here.

Monday, 7 March 2022 – Steiner Ranch, North Carolina, USA

5 p.m.

I woke up to a few text messages from Pietro's father this morning. He, like his son, is very disappointed that he didn't get the seat and he's voiced his opinion to me strongly. It's OK. I'm used to it and I completely understand.

Anyway, forget my comment yesterday about the future starting here. Things have gone from shit, to really shit, to positive again with Kevin, and now back to shit again. The Bahrain test starts on

Thursday, 10 March and I was informed late last night that our cargo was still stranded in the UK. The transporter plane that was supposed to be flying it there got delayed at Istanbul airport. Worse still, we have no idea exactly when the cargo will leave. Every other team will be setting up for the test now, whereas at this rate we'll be lucky to start setting up by foking Wednesday! It is a very serious situation and I have a meeting in fifteen minutes to see where we are.

Somebody suggested to me yesterday that the only reason Haas are in existence is to create content for Netflix and I'm beginning to think they are right! What is it with us? Have I done something evil in a past life? If I had, at least it would explain to me why we are having so much bad luck at the moment. In Barcelona we manage fewer laps than any other team and through no fault of our own. Then, thanks to Putin, we become the most controversial sporting team on the foking planet and have to get rid of a driver and our title sponsor. Then, as if that isn't enough, the plane that is supposed to take our cars and the majority of the parts from the UK to Bahrain breaks down in Turkey of all places, putting us two days behind schedule. Somebody please tell me, what the hell have we done?

Tuesday, 8 March 2022 – Steiner Ranch, North Carolina, USA

7 a.m.

I'm meant to be catching a flight to Bahrain very soon and it looks like I'm going to arrive there before our cars! The situation still hasn't been resolved. The last thing I heard was we were trying to arrange another cargo aircraft from East Midlands Airport to get the freight to Bahrain via Leipzig. Leipzig!? At the moment there is a serious chance that we will have to write off the first day of testing, which frankly would be a disaster.

9 p.m. – Team Hotel, Bahrain

Unfortunately I have beaten our cars to Bahrain. Then again, they should be landing any moment now, so if the team pull out all the stops, which they will, there's a chance that we'll be able to get out for the second session on day one. If we do, it'll be a miracle.

I must admit it's been hard to remain positive these past few days. I know I joked that we must be cursed or something, but when shit like this keeps happening to your team you begin to wonder if your luck is ever going to change. I know I have a reputation for having a laugh and a joke but that isn't why I'm here. I'm not here to be made fun of, either. I'm here because I want to compete, and compete at the very highest level. That's what I strive to do every single day of my life. Ask anyone who knows me and they'll tell you the same. It's what I wake up for.

In the end I had to have a word with myself on the flight over here and remind myself that there are positives in the situation. We're here, which is the main thing. Here in Formula 1. And here in Bahrain. Finally! That in itself is a big achievement after what we've all been through since 2020. And we've got a new car, of course. Two years in the making and hopefully worthy of some points this time. Or maybe even a podium. Who knows? I said earlier that the atmosphere that hope creates is the lifeblood of every sport. I still think that's the case but in order to create that hope in the first place you need the right people behind you. For the first time in years, I believe that we have not only the right people, but the right infrastructure in place to create the hope that we need in order not just to survive as a team, but also to thrive. I could have said 'drive' there to match the title of the book, but never mind. We also have a greater level of technical staff at Haas than we've ever had previously, which is all thanks to Gene. The team has never been as big as it is now and hopefully that will start bearing some fruit. So what am I complaining about? Come on, Guenther.

TESTING

Thursday, 10 March 2022 – Bahrain International Circuit, Sakhir, Bahrain

9 a.m.

As if proving my point from yesterday, the team here in Bahrain have worked their asses off for the last thirty-six hours and it looks like we'll be good to go for the afternoon session. OK, so we've lost four hours. But from where we were on Tuesday, we've actually gained a day. I couldn't be more proud of these guys. I'm so ready for this now. We're back!

Kevin is coming in for his seat fitting very soon. It's all happened very last minute but he's here and that's obviously the main thing. It's his first day in the car tomorrow and everybody's really excited. I can't wait to see his little Danish face again.

One or two people have been asking me if we considered Romain for the job and the truth is we never did. He and Kevin are on a different path, I think, and the one that Romain has taken, which involved moving his family to Miami and signing a multi-year contract with Michael Andretti for IndyCar, ruled him out. He was a man with a plan, you know, and is also five or six years older than Kevin. Kevin's more like me. If he sees a challenge, he grabs it.

5 p.m.

Pietro was in the car today. He's a good guy. Some drivers would have just sulked their asses off after not getting the drive but he was all smiles and ready for action. Everybody likes him. Because of what happened in Barcelona we decided to run with harder compounds and longer lap simulations in order to try to maximize the number of laps. All in all, it was quite a good day of running. We're getting a better understanding of how the car works and we just need to continue like this. Forty-seven laps in total. Not bad but not brilliant. We need more.

Friday, 11 March 2022 – Bahrain International Circuit, Sakhir, Bahrain

4 p.m.

I got a really good night's sleep. The kick up the ass I gave myself the other day has worked and I'm back to firing on all cylinders. All fifty of them! I also like Bahrain. The people are nice, the hotel is comfortable and I like the track. Kevin arrived in the garage bright and early this morning. It was his first time in the Haas garage since December 2020 and it was great to see him. You know, I remember the day when Kevin first arrived at our headquarters in Kannapolis. This is before I got rid of him, by the way. It was 26 January 2017 and for a laugh we got him cleaning and doing the photocopying. I can't think of the last time the entire garage was smiling like that. Already the effect of him coming back has been amazing.

Anyway, to business.

Mick was first out in the morning. His departure was delayed by almost an hour, though, with an oil leak. Great! After that we had a problem with the cooling system and then an exhaust issue, so he only managed to bank twenty-three laps. Aaaah! How foking frustrating. Fortunately, Kevin fared better in the afternoon. The last time he sat in an F1 car was at the Abu Dhabi Grand Prix in December 2020 but it was as though he'd never been away. He did sixty laps in total.

Because of the delay to us starting, we've been granted an extra four hours over the test and so we continued for an extra hour today. We could have done with an extra day, though. We still have a lot of gremlins in the car and it's starting to worry me. In total we've lost about three days out of five over the two tests. That's a lifetime in this situation.

After the session I asked Kevin how he felt. 'I think my neck is broken,' he said. 'I'll do another day tomorrow, though, break it even more, and then hopefully get in a little better shape for next week.' That's typical Magnussen.

NEW SEASON

Saturday, 12 March 2022 – Bahrain International Circuit, Sakhir, Bahrain

9 p.m.

What a difference twenty-four hours makes.

We kicked off early at 9 a.m., so exactly one hour ahead of the rest of the field. I won't name them here but some people on the grid were actually against us being granted some extra time. Can you believe that? Formula 1 are responsible for the freight and the issue with the plane was nobody's fault. These people who are complaining are just idiots.

Kevin quickly added to the sixty laps he recorded yesterday afternoon and went straight to the top of the time sheets. It wasn't a problem-free morning, though, a fuel system issue ending his run earlier than planned. He managed to get a few more laps in before the session ended, and finished on thirty-eight laps with a best lap of 1:38.616.

After lunch, Mick arrived to close out the test. Yesterday was poor for him through no fault of his own, so we needed some luck. This time we had an extra two hours at the end of the session, which was under the lights. In addition to completing eighty-five laps, which was encouraging, he ended up setting the second-fastest time of the day with a time of 1:32.241.

So that's it, then. No more testing. I think we had just over three days in total out of a possible six, so about 60 per cent. Four or five days ago I would probably have sunk into a pit of frustration at that, but not any more. I am only interested in positives now. The car is looking promising, the drivers are happy, and we're progressing as a team. Apart from more testing and a hundred million dollars a year more to spend than the other teams, I couldn't really ask for more at the moment.

What we need to start working on is the reliability. We didn't have it at any time during the tests and we've got a big mountain to climb. Performance-wise, it's difficult to say because we haven't done enough running, but I would say it doesn't look bad.

Somebody asked me a few hours ago if I was surprised that Kevin had integrated back into the team as quickly as he has done. I suppose I am a little bit. Then again, that was one of the reasons why we brought him back. He knows the team and he's a good driver. Obviously, I was happy with his performance during the test but, once again, why would that surprise me?

I think the driver line-up we have now for 2022 is a strong one. It has youth and experience and I think that, as the season progresses, the drivers will start to complement each other. One of Kevin's big strengths is that he's been through a lot of ups and downs in his career, which has made him tough. He's well prepared for the challenge ahead and has been there and has bought the foking T-shirt. Mick is someone who wants success and will do whatever it takes to get it. His immediate challenge, in my eyes, is to be nipping at Kevin's heels. If he can do that I will be very impressed. Kevin will bring Mick on, I'm sure of it.

The person who asked me about Kevin also asked me what would represent a success for Haas at the Bahrain Grand Prix. These sorts of question usually get on my nerves a bit, and especially at the start of the season. That said, I have to admit that since yesterday I have been daring to wonder how we might get on. It's what happens when you think you have a good car. And I think we do.

'First of all I'd like to finish the foking race,' I said.

That's the truth, though. The reliability issues are our biggest headache right now so if we can stay on the track for the whole race, not only will that be an achievement but it will leave us in with a chance. I'm optimistic. Cautiously so, but optimistic. Actually, balls to being cautious. I think we'll end in the points.

Friday, 18 March 2022 – Bahrain International Circuit, Sakhir, Bahrain

8 a.m.

Some people have butterflies in their stomach when they arrive for the first Grand Prix of the season. I have foking eagles! I'm being serious. I managed to sleep OK but as soon as I woke up they were there, circling and flying about. This is why we do what we do. I love the challenges and opportunities that testing gives us and it's great to be at a track. You can't beat the atmosphere of a Grand Prix, though.

As I said before, I like the Bahrain circuit. Historically we've had some success here and we know the track well. Kevin alone has raced here seven times and in 2018 he finished fifth. Mick too won the 2020 Formula 2 Championship in Bahrain and also made his Formula 1 race debut here. I still believe what I said yesterday. We're capable of scoring points.

8 p.m.

We had a very clear plan in FP1, which was to go heavy on fuel, do as many laps as we could and basically carry on testing. There are quite a few things we need to find out about the car that we couldn't do here last week so it made sense. As a consequence, we finished eighteenth and nineteenth, which gave the naysayers an opportunity to say, 'Oh, I see Haas are still where they belong.'

Fortunately, we found out most of what we needed to know, which meant in FP2 we could go out with the same fuel loads and engine modes as the other teams. Also, the time of day and cooler temperatures were more reflective of what we'll all face in qualifying and the race. Mick recorded a 1:33.085, putting him in P8, and Kevin in P10 with a 1:33.183. Before the session ended, they both did a couple of high-fuel runs and that was that.

All in all it was a good day. There were no issues with the car. No gremlins, nothing. The team have done a fantastic job over the last week – from testing to now. It's been relentless. We just need to carry it over to qualifying with the same pace and reliability. We can do it.

Saturday, 19 March 2022 – Bahrain International Circuit, Sakhir, Bahrain

8 a.m.

The first person I saw when I arrived in the paddock this morning was Fred Vasseur from Alfa Romeo. 'Oh foking hell,' I said to him. 'This is getting boring now. Are you stalking me? You are, you're stalking me. Just leave me alone, you French idiot.' Fred and I have been neighbours in the pit lane for a while now and out of all the team principals he's the one I get on with the best. Well, him and Mattia. I actually get on pretty well with all the team principals at the moment. We might all be competitors but it's within a sport that we're all very passionate about. As well as stopping us from killing each other, it's probably what holds us all together. We also have to do media together sometimes so if I wasn't on speaking terms with anybody it would be pretty uncomfortable. Life's too short for shit like that. It doesn't make much of a difference with Fred because he can hardly speak anyway. Maybe that's why we get on.

Everybody was in good spirits when I arrived just now. Not just in the team but in the paddock, too. Sometimes it's like a morgue. Especially if the weather is shit. Fortunately we don't have that problem here and because it's the first race of the season we're all happy. For now! You wait until we come back after qualifying. One or two people in the paddock will have a face like a smacked arse then, I guarantee it. Hopefully not me.

6 p.m.

Wow! Where do I start? To give you a clue how it went for us this afternoon, my face is smiling.

The idea in Q1 was to qualify for Q2 in as few laps as possible. 'Let's not take any chances,' we said. We were prepared to do more if we needed to but that was the idea. Free Practice went well (Kevin finished seventh overall and Mick fourteenth) but during that session Kevin's car developed an oil leak in the hydraulic system. If we'd tried to fix it we wouldn't have made qualifying so we had no choice but to refill the system between runs. To do this you have to remove the panels, which takes up valuable seconds and, just to make matters worse, because of what was happening, Kevin kept on being called into the weighing station. Despite all of this both drivers managed to get through Q1 quite easily. 'Jeezoz Christ,' I said to Ayao. 'That's the first time this has happened since 2019!' It was a shock to the system. We'd become so used to struggling. I know it's not a true reflection of the potential on the grid, but Kevin finished fifth in that session.

In Q2 Kevin continued where he'd left off and finished seventh with a time of 1:31.461. Just remember, until last week this guy hadn't even sat in an F1 car for over fourteen months. So what does he go and do? He gives us our first appearance in Q3 in three years. Incredible! Unfortunately, Mick made one or two mistakes and wasn't able to match Kevin's success. He finished twelfth, though, which isn't too bad. He made Q2 for the first time in his career. Next time, Mick, how about Q3? It appears the car is good for it.

Because of Kevin's oil issue we knew that we could only do

one lap in Q3. Even if we'd had two sets of tyres left it wouldn't have made any difference. In the end we waited until the very end of the session before sending him out. This was actually for the benefit of the other drivers – if Kevin had broken down on the track at any other time during the session the yellow flag would have got us all killed. A couple of cars did end up following us out onto the track but we tried our best to go as late as we could.

Unfortunately Kevin could smell the oil as he was driving. Bearing in mind what could have happened if it had made it on to his rear tyres, this obviously made him nervous. As a consequence, he probably didn't go as fast as he could have but he still qualified seventh, which was obviously amazing.

Not a bad start to the new season, then. Let's see what happens tomorrow.

Monday, 21 March 2022 – Bahrain International Circuit, Sakhir, Bahrain

2 a.m.

Fifth!! Kevin finished fifth. Last season we were the only constructor not to score a single damn point and we have started off the 2022 season with ten. Ten points! It's 2 a.m. here and I am dead on my feet. Happy, but so foking dead. I need to go to bed.

My phone is still ringing and I must have had two hundred text and WhatsApp messages since the end of the race. Some people seem to be surprised that we've scored points but they all have very short memories. I'm relieved, but I'm not surprised. Even Gene was quite animated when I spoke to him. Usually he's very impassive, but not today.

Somebody asked me last week if I can make Gene Haas laugh. Yes, I can. Not because he necessarily thinks I am a funny guy. I always come out with so much shit, though, and sometimes he just sits there laughing in disbelief. And you should see the board meetings! These people are all Californians and when I start talking they just sit there with their mouths open. And I can tell what they are thinking. They are thinking, *Jeezoz Christ, did he really just say that?*

Everyone was very nervous this morning. I've never known the garage to be so quiet. Usually the mechanics are telling jokes and taking the piss out of each other – and out of me, mostly – but this morning it was different, you know? The calm before the storm. Everybody said good morning but that was it. In the end I called the team together and said, 'Look, you guys, the reason we're all nervous is because we've got a foking good car for a change and we're competitive. I know you believe in each other so start believing in the car! It's going to be good, OK?' The team seemed to brighten up after that and we got a bit of energy going. I might have to change my name to 'The Incredible Bullshitting Man', though, because at the time I was shitting myself. Perhaps I am a good actor after all? It wouldn't surprise me. After all, I managed to talk the FIA into giving us a licence all those years ago.

One of the decisions that Gene and I have made for this season that we knew for sure would work out was asking Kevin to come back. I don't regret getting rid of him and Romain in 2020 and I've already explained why we did it. Taking on a pay driver worked for us at the time and we've all come back to this a lot bigger and a lot stronger. Fok, it's good to have him back, though. Who knows, if

Mick pisses me off maybe I'll call Romain? That's a joke, by the way. Mick's going to be OK, I think. At least I hope he is. There's certainly a lot of promise there.

What do I remember about the race? God, let me think. Reliability was still a worry and shortly before the race Ayao made the mistake of reminding me that the highest number of laps we'd done in the car in one run was eighteen.

'And how many laps is the race?' I said.

'Fifty-seven,' said Ayao.

Thank you very foking much!

Kevin had a great start and worked his way up to fifth on the opening lap, which was just incredible. Mick also had a very good start and moved up to ninth. I remember saying to Ayao on the pit wall, 'What the fok is happening here? Everything's going to turn to shit, you watch!' He just smiled. It's best not to encourage me when I'm being pessimistic and gloomy. Just ignore me.

Kevin pitted for softs on lap fourteen, and again on lap thirty-four for mediums, before coming in once more for softs on lap forty-seven when a safety car was deployed. Before that Mick had been tapped into a spin by Ocon from Alpine, who received a penalty, and by the time the safety car came off the track Mick was back thirteenth. He was quick, though, so we thought he might still make the points. 'You've gone quiet again, Guenther,' I remember Ayao saying. I didn't reply. I was now feeling confident for some reason and when I feel confident I also get nervous. Why? Because I'm so used to everything going foking wrong!

Kevin held on to seventh at the restart and then on lap fifty-four Verstappen retired due to a fuel pressure problem and Perez for the same issue two laps later. And where were they positioned? Fifth and sixth, which moved Kevin up to fifth. He seemed to be under pressure from Bottas but was holding him off well. I must have lost my mind here for a second because I had to ask Ayao how many laps were left. 'One, you idiot,' he said. 'You know that!' I did know that but I just couldn't think. We were one lap from finishing fifth in the opening race of the season.

What seemed like an hour later I heard Kevin's engineer over the radio. 'That's P5, mate,' he said. 'Well done.' It was then that I came alive, like Frankenstein's monster!

I don't remember any of this but apparently my first words to Kevin over the radio were, 'Kevin, that was some foking Viking comeback! I cannot believe it.' Kevin couldn't hear what I was saying because everyone was screaming in the background so I don't remember saying it and he didn't hear it – yet everybody's saying it!

What a crazy day.

The difference in the team has been immediate. For the last two years, every time we've had a double or triple header the mood in the team after the first race has been shit. Everybody just wanted to go home. Next week we're in Jeddah and, just for a change, we can't wait to get there. We have energy because we have hope. I'd almost forgotten what this was like. Unfortunately, the team can't really celebrate as we have to pack everything up for the morning. We'll make up for it, though, somewhere along the way.

One of the journalists after the race asked me where this result fits in my top ten moments at Haas. 'Surely it must be up there at the top,' he said. It was actually a good question and before I answered it I had to think about it.

I know I've mentioned it lots of times so far but what makes the result special is the fact that it's happened in a comeback situation.

Everything else on the list of my favourite moments at Haas, like finishing fifth in the Constructors' Championship, happened as part of a progression, so although we didn't take any of it for granted, it was what we were working towards. The result today has come off the back of two years of shit and because it's a new car that has been built by a new team in a new facility, we've had nothing to compare it to. Or at least nothing worth talking about! It's been like starting again, almost. Or, at least, that's what it's felt like.

It's no secret that Gene was thinking about walking away when he reviewed his investment and, despite what was happening with Covid, he could have sold the team in a second. Whether he would have made all his money back I'm not sure but he could have walked away very easily. He didn't, though. And why didn't he? Because he, like me, believed that we could achieve what we have achieved today and more. Gene Haas is no fool, you know. He's a clever guy and if he didn't believe in us he'd have pulled the plug.

I'll tell you what else feels good about this and that's proving the naysayers wrong. In 2016, when we started, they all said that being a customer team and buying all the parts from Ferrari wouldn't work. Then what happens? We go and score points in three out of the first four races. Whether it's the fact that I'm a bit of a joker and am not corporate or conventional I don't know, but some of the same naysayers have been condemning us ever since then and have been willing us to fail. Not everybody likes small teams, you know. We're still here, though, and as much as that might piss some people off, we're still relevant. Today proved that.

Do you know the most satisfying thing to happen to me today? This is even better than greeting Kevin when he arrived back in the garage, which was pretty foking special. It was being approached by a couple of team members just before I left for the hotel tonight. 'You told us that we'd be back in 2022, Guenther,' they said. 'And you were right. Thank you.' They meant it, too. I could tell. That was amazing.

Thinking about it, that is probably at the top of my all-time favourite moments at Haas so far. I knew that when I made those speeches and talked to the staff that some of them wouldn't believe me. That's completely understandable. Formula 1 was almost on its knees and the whole world was sitting under a cloud of uncertainty. 'Yes, but Guenther Steiner says it's going to be OK!' 'That's all right, then!'

Now, to bed. I'm foking exhausted.

Thursday, 24 March 2022 – Jeddah Corniche Circuit, Jeddah, Saudi Arabia

10 a.m.

I arrived in Jeddah this morning and went straight to the track. The second race of a double header (or a triple header, if you include the test) and I'm feeling good. We all are. The last time we were here in 2021 there was shit running down the back of the garage walls. I'm not kidding! The track had been completed in record time and there was a problem with the sewage system. I'm sure it will have been sorted out by now. Otherwise, we're going to smell like a toilet. Anyway, the guys are just about set up now so we're all ready for tomorrow. Bring it on!

8 p.m.

I've been doing a lot of press since Bahrain and that's been fun for a change. Over the past two years almost every conversation I've had with a journalist has started off with something like, 'So, Guenther, another disappointing day for Haas. How do you feel right now?' What I actually want to do at that point is say: 'How do I feel? How do you *think* I feel? I feel like shit!' Then just walk away, sit down and have a coffee. But you have to play the game. 'Yes, well, you know things are difficult at the moment, blah blah blah.' It's the same old shit but I understand that it has to be done. A necessary evil.

Kevin too has been doing a lot of press just lately and I've been reading some of his interviews. He's a funny guy for a Dane. Dry as a bone. I'm paraphrasing now but in one of the interviews, when he was asked about how he came back to Haas, he said something like: 'I went away after Guenther sacked me and built myself a really good career. I grabbed podiums, pole positions and even a win. I was really, really happy and I was enjoying it. But then Guenther called me again and ruined all that . . .'

When I read that I almost foking wet myself. What a journey, though. We've gone from scoring a shitload of points together to him leaving for being useless and then coming back again for being amazing and not Russian!

It's too early to say for sure but it looks like him and Mick are going to have a really good relationship. I think Mick realizes that, as well as being easy to get on with, Kevin has a lot of experience and can mentor him. If he allows him to. There aren't many drivers on the grid with as much experience as Kevin so he should really grab the opportunity.

I had a long conversation with Gene last night. Calling him at the moment is strange because I'm not dreading it! I can't remember the last time it was like that between us. For the last two years my version of giving him good news has been telling him we finished seventeenth and eighteenth instead of nineteenth and twentieth. I'm telling you, that's the truth. When my phone goes and I see his name my heart sinks, and I'm sure it's been the same for him. 'Oh shit! It's Guenther. What's happened now?' This last week has been a lot more relaxed and, as opposed to just calling each other back, which is what we do sometimes, we pick up immediately. We've only scored ten points, so imagine what will happen if we get a podium. We'll probably buy a house together!

So, what am I hoping for this weekend? Apart from talking about Bahrain, which I'm still happy to do at the moment. That's the thing I'm being asked the most by the press. It certainly makes a change from what I'm used to! 'What are you expecting this weekend, then, Guenther? Eighteenth, nineteenth, or a DNF?'

Well, I'd be lying if I said that I wasn't expecting to be in the points again. This time we want it to be both cars, though. Not just one.

Over to you, Kevin and Mick.

Friday, 25 March 2022 – Jeddah Corniche Circuit, Jeddah, Saudi Arabia

10 p.m.

The reason I'm still at the track is because a missile strike took place earlier about 10 miles from the circuit. Apparently the target was the Aramco oil refinery. I feel like a foking war correspondent! As you would expect, everybody on the grid is worried and there are meetings going on all the time with the organizers, who are updating us on the situation. What everybody wants to know is how safe we all are because, if we aren't, we can't possibly stay in Jeddah. If that's the case, we'll be out of here as quickly as we can and there will be no Saudi Arabia Grand Prix. I've never known a situation like this before.

The last time a Grand Prix was cancelled because of something other than Covid was in Bahrain in 2011. I was working in Nascar at the time but I remember it as clear as day. About a month before the race was due to take place, some anti-government protests started taking place that became known as the Arab Spring. Eventually they reached Bahrain and during the protests several people were killed. It was feared that the protesters might try and exploit the Grand Prix in some way because of its global reach and so a week or so before it was to take place the crown prince of Bahrain, Salman bin Hamad Al Khalifa, called the race off.

I know what you are thinking now. You're thinking, 'Guenther has just looked that up on the foking internet.' OK, I had to look up the details. I admit that. I remember it happening, though.

In about fifteen minutes we have a meeting with the minister and the organizers so fingers crossed.

So, what about the non-missile part of the day? FP1 was a mixed bag really. Kevin was limited to just two installation laps before *another* hydraulic leak forced him to remain in the garage for repairs. Mick fared better with twenty-two laps, so all in all not too bad. Where is the reliability, though?

After that I did a couple of interviews and the subject of Nikita

Mazepin came up. Since sacking him and severing ties with Uralkali, we've remained pretty tight lipped. Partly because we have nothing really to say on the matter but mainly because we've got other things to think about. Like a Formula 1 season! A couple of weeks ago, Nikita and his father were included on a list of people who now face sanctions from the European Union over Russia's invasion of Ukraine. I've just looked it up now and Nikita's entry on the list, which was released by the European Council on 9 March 2022, says:

Nikita Mazepin is the son of Dmitry Arkadievich Mazepin, General Director of JSC UCC Uralchem. As Uralchem sponsors Haas F1 Team, Dmitry Mazepin is the major sponsor of his son's activities at Haas F1 Team. He is a natural person associated with a leading businessperson [his father] involved in economic sectors providing a substantial source of revenue to the Government of the Russian Federation, which is responsible for the annexation of Crimea and the destabilization of Ukraine.

The only thing they have wrong with that is that it should be in the past tense. Uralchem *was* a sponsor with the Haas F1 Team. It certainly isn't now. The reason I mention it is because on the day the list was released, Nikita made a statement saying that he intended to force his way back into Formula 1.

Unfortunately, FP2 was a disaster for Kevin. He was forced to retire with more mechanical issues and without even clocking a flying lap. Shit! The positive is that Mick had another clear run and logged twenty-seven laps, before finishing off with a high-fuel run.

I was about to say what an eventful day, but it isn't over yet!

Saturday, 26 March 2022 – Jeddah Corniche Circuit, Jeddah, Saudi Arabia

8 a.m.

Good news. The Grand Prix is on!

Shit, am I tired, though. I don't think I got to bed until about 2 a.m. and it took me a while to get to sleep. Missile strikes tend to have that effect on people, you know.

The outcome of the meetings (there ended up being several in the end) was that we came away feeling assured that everybody would be safe and so the decision was made to go ahead. I am personally responsible for the entire Haas team and if there had been any doubt at all in my mind I would have moved them out and sent everybody home. I see no issue, though. It's all good.

OK, now we prepare for qualifying.

5 p.m.

Mick had a big shunt during Q2 and is in hospital. He's OK, though. As far as I know he went too fast on to a kerb at Turn 12, lost the rear end of the car and went straight into the barrier. We lost contact with him on impact but the medics were there within seconds and he was taken to the trackside medical centre. After that he was transferred to the nearest hospital in Jeddah for precautionary checks and I've now been given an assurance that he's OK. It obviously brought back memories of what happened with Romain back in 2020. That was totally horrific and I think it haunts us all to a certain extent.

What a foking weekend, though. And we haven't even had the race yet! So far I've to deal with ongoing reliability issues, which is a pain in the ass, people asking me about Mazepin, who was an even bigger pain in the ass, getting attacked by missiles, which could have been a deadly pain in the ass, and then Mick crashing into a barrier at 170 mph, which was just terrifying!

Qualifying actually started well for us. Both drivers made it into Q2, as they had done a week earlier, and everyone was happy. Q2 was then marred by Mick's crash, which resulted in a significant red flag delay. He'd just posted a good time on a set of softs and was heading out with a new set on when his accident took place. Kevin finished P8 so made it into Q3 for the second successive qualifying. His first flying run was his quickest and he finished in P10. All things considered, it was a good effort really.

Anyway, forget qualifying for a moment. The dilemma I had when Mick had the crash was whether or not to ask the mechanics to work through the night and make a new car for Mick or withdraw the car from the race altogether. In the end it only took me about two minutes to decide and I've already withdrawn the car from the race. For a start, I didn't know how Mick was until about half an hour after the crash and that had a bearing on things. Also, in addition to me having to ask the mechanics to work through the night after already having coped with testing and then the Bahrain Grand Prix, had something gone wrong during the race we might have compromised our chances in Australia. Something like that would have put everyone and everything on a downward spiral and I couldn't allow it. Mick would also have been starting the race from the pit lane, so the chances of him scoring points would have been minimal. It's a similar situation to us not developing the car in 2021 and putting all of our attentions on 2022. Not as much is at stake, of course, but it's the same principle.

Despite the crash and everything else, at the moment we're in a good place. We just need to be patient, work constructively and wait for the points. In the end Mick was released from hospital very quickly and could have raced. The decision had been made, though.

OK, race day tomorrow. Given what has already happened, I honestly don't know what to expect.

Sunday, 27 March 2022 – Jeddah Corniche Circuit, Jeddah, Saudi Arabia

8 a.m.

Fortunately, I had a much better sleep last night. Also, when I switched my phone on earlier, there were no messages about disasters. Amazing! I just have to do one telephone interview after breakfast and then I'll be off to the track.

5 p.m.

We scored points! Only two, but it's better than nothing. I can't remember the last time we scored points at consecutive races. It's a good effort, and especially after everything else that's happened this weekend. We just have to keep it up now and try and get Mick involved.

We started on a different strategy to most of the other teams (we were on a hard tyre and everyone else was on mediums) and unfortunately the safety car came out just at the wrong moment on lap sixteen after an incident involving Latifi. Kevin eventually pitted but when he came out he was down in twelfth. We did get a little bit of luck towards the end of the race with the VSC, which helped us move up into the points. The frustrating thing is that had that been the real safety car, everyone would have been bunched up together with new tyres on and we would probably have scored big again. Anyway, it was a good race for Kevin and he said the car was phenomenal today. That makes me almost as happy as the two points. His neck is pretty foked, though. Never mind.

I did think earlier that perhaps I should be talking more about the Grand Prix in general and about the winners and the title race. Then again, that would mean me talking about all the other teams, drivers and team principals, and I couldn't give a shit about them. It's Haas all the way, baby.

Laters!

Monday, 28 March 2022 – Team Hotel, Jeddah, Saudi Arabia

8 a.m.

I slept well again, which is good. No surprise really. I hope not every weekend is like this one. I don't think my heart could take it.

I've just been given a slap on the wrist from Stuart, although only a joke one. I did an interview with a satellite radio station after the race yesterday and they suggested to me that if it had not been for the yellow flag we could have had more points. I replied that we always want to score maximum points but that we couldn't be too greedy as last year I would have hugged the entire paddock for two points. Apparently, they thought I said 'foked the entire paddock' (me?) and they had to make an apology. I don't blame them for thinking that. Remember, I learned my English in a rally garage, guys. If you want a team principal who definitely won't swear, try . . . actually, I can't think of anybody. They're not all as bad as me but they're all capable of swearing.

I've got a flight back to the States soon. I go from Jeddah to New York and then New York to Charlotte. About twenty hours door to door, I think. It's going to be good to see Gertie and Greta, but do you know what I'm also looking forward to? The conversation when I walk through the door. Have you any idea what it's like going home, week after week, when you're doing shit? Instead of giving you a hug when you arrive, they pass you a box of tissues. 'Oh God, Haas came last again. Make sure you're nice to Dad when he gets home.' This time when I arrive I'm going to foking smash the front door down and shout, 'TWELVE FOKING POINTS!'

Mind you, if I did that I'd be living in a hotel.

Friday, 1 April 2022 – Charlotte Airport, Charlotte, USA

8 a.m.

The whole *Drive to Survive* juggernaut has really started to kick in now. I don't know exactly how long the new season has been out for but ever since I arrived back home the selfies and conversations have been almost constant. I don't mind at all, you know. Everybody's very respectful, so if I'm out with my family or I'm obviously busy they usually leave me alone. Everybody wants to know about Nikita. I'm very diplomatic and say that I couldn't possibly comment. Well, most of the time. They also like quoting things that I have said in the show like, 'We look like a bunch of wankers.' That seems to have become my catchphrase and I don't remember even saying it! 'Are you sure I said that?' I reply. 'I never usually say bad things.'

The other part they like to talk about is when I lost my shit with Kevin when he smashed my door. Now that I *do* foking remember. I had given him and Romain a bollocking about something and as they were leaving the office Kevin smashed my door. The little Viking shit. I should have made him pay for it.

So, what have I got to look forward to today? What, you mean apart from a thirty-hour flight to Melbourne? This is one of the journeys that seems to result in people getting really bad jetlag. Personally, I don't suffer from it too badly usually, so I suppose I'm lucky. I'm sometimes asked how I do this and I just say that wherever I am in the world – it doesn't matter where – I get up when the sun rises and I go to bed when it sets. That's actually bullshit but it sounds good. Gene suffers quite badly from jetlag, I think. Whenever I see him he's always tired.

I have never attempted to add up the miles I travel every year but before the end of the season I will try. It might be similar to the swear words I say in a week but I don't know anyone brave enough to try and count them. You'd need a big foking calculator, put it that way!

I'm really looking forward to arriving in Australia. Everybody loves going there. The weather normally is good, the temperature's fine, the people are just amazing and it's a good track. Actually, the people aren't just amazing. They're foking crazy! But in a really good way. I've never known anyone like the Australians. Our best result there so far is sixth, which Kevin managed in 2019 and Romain at our very first race in 2016.

Right, my car is here. I'll have a think about the miles I do on the plane.

Sunday, 3 April 2022 – Team Hotel, Melbourne, Australia

10 a.m.

It's about 80,000 to 90,000 miles, I think. That includes the trips to see Gene and the trips to the factories. That's about three and a half times around the Earth. Fok me!

We landed in Melbourne at 6 a.m. To be honest I haven't been at my best these last few days and because of that I have a feeling that the jetlag might have more of an effect than usual. I hope I'm wrong but I didn't manage to sleep at all on the flight, which is unusual for me. Anyway, let's see what happens. I'm still standing. Just!

Once again, *Drive to Survive* is never too far away. I'd only been off the plane five minutes when five Aussie guys came up to me and asked for a selfie. Australians are always so enthusiastic about things. They're great people. Sometimes they get too excited, though, and when that happens I cannot understand them. I just have to smile and nod my head.

Once again, the main topic of conversation when I met these people was Nikita. 'He seems like a right spoilt little bastard!' one of them said. Having not watched the show, I never know exactly what they're talking about, but one thing's for sure, it's not very positive for him! I've also had the press asking me about the Mazepins. Something about a leaked letter and them wanting a refund on their sponsorship money. I keep on telling the press that I have no comment to make and that will never change. Gene and I made the decision to sever ties with them both and that's where I stepped away.

Tuesday, 5 April 2022 – Albert Park Circuit, Melbourne, Australia

2 p.m.

It doesn't matter how many times I come here, Albert Park will always remind me of our first ever race back in 2016. We could have started racing in 2015 but decided somewhere along the way that we weren't quite ready. That was a difficult one to swallow, as everyone wanted to get started, but it was the right thing to do. Just as coasting in 2021 was the right thing to do. I should really tell you about that race because it's an important day in our history. In my history!

Our drivers were Romain and the Mexican driver Esteban Gutiérrez, who had been at Sauber the year before. He only lasted one season with us and was replaced by Kevin. Had we started racing in 2015 our expectations would probably have been zero but that extra year of preparation and development gave us hope. We were also the new boys in town and so the main emotion for us was excitement, really. I remember it rained during FP1 and Gutiérrez only managed eight laps and Romain six. FP2 wasn't much better but it was the same for everybody. If my memory is correct Gutiérrez finished the session in tenth and Romain thirteenth, which was foking OK, you know.

At the start of FP3 I got my first taste of what Romain was really capable of when he collided with one of the Manors while exiting the garage. He was straight back in for a new floor and had travelled about 4 metres! Although both drivers managed more laps, they posted shit times and were right at the back.

What this Grand Prix is famous for, apart from the debut of Haas F1, is the knockout-style qualifying format, where the slowest driver was eliminated every ninety seconds. You remember that? The idea was to force the drivers to be continually fast but it turned out to be a bit of a shit show as there were long periods where there were hardly any cars on the track. Gutiérrez was the third driver to be eliminated after the two Manor drivers, and Romain ended up being the fourth, which put us nineteenth and twentieth on the grid. Nobody liked the format. Least of all me! It was completely shit.

Race day was dry and after Kvyat, who was starting eighteenth, broke down on the formation lap, both our drivers gained a place before we'd even started. He's my favourite Russian! By lap eleven Romain was in fourteenth and Gutiérrez sixteenth. On lap seventeen, Alonso, who was driving for McLaren, went to pass Gutiérrez and clipped him on the inside, sending both drivers into the gravel.

Alonso's crash was pretty spectacular and ended up in an airborne barrel roll. He was OK, though, apart from some cracked ribs. Gutiérrez was fine but his car was obviously foked. Welcome to Formula 1! A red flag occurred a few laps later and because Romain had not pitted yet he was given a free stop. On lap twenty-two, Räikkönen retired, which put Romain in eighth. There was still a long way to go but I remember feeling both anxious and excited at the same time. Actually, who am I kidding? I was foking crapping myself!

I have to give credit to Romain here because the thing that gave him an advantage in the closing stages of the race was that he had taken good care of his tyres and by the end of the race he'd managed to gain two more places. 'We finished sixth,' I said to Gertie afterwards. 'And in our first race!'

'Where did the other car finish?' she said.

'Don't spoil it!' I said. 'We finished sixth. That's all you need to know.'

The drivers who finished ahead of Romain were Massa, Ricciardo, Vettel, Hamilton and Rosberg. So, the best drivers on the grid. He got driver of the weekend, which he deserved. More importantly, what happened that year with us in Melbourne is credited as being one of the strongest debuts by any team in the history of Formula 1. Anyway, enough of the bragging already! That was then, this is now, and we have work to do.

What a great weekend, though!

Wednesday, 6 April 2022 – Albert Park Circuit, Melbourne, Australia

Kevin only arrived in Australia this morning and he looks like shit. I'm not feeling too good, either, but he looks like a ghost. It's pretty worrying.

The Las Vegas Grand Prix for 2023 has now been confirmed. So far, I haven't spoken to anybody who doesn't think that it's a good idea and I'm really looking forward to it. Ever since *Drive to Survive* started, F1's following in the States has increased massively year on year and so the timing is perfect. Obviously, we've got Miami coming up this year too, so it'll be three races next year. Do I think we could have four races one day in the States? If things keep growing at the pace they are at the moment, why not?

I think Las Vegas fits well with Formula 1. Glamour, entertainment and bullshit. That's what both are all about. Some people will say that we are doing too many races now but I don't think that's the case at the moment. I think Stefano has said that twenty-four Grands Prix would be the absolute limit, which sounds about right. In 1980 there were just fourteen Grands Prix. Can you believe that? One of the other advantages of holding Grands Prix in new locations is that it should keep the organizers of the more historic races on their toes. Nothing is certain in this sport and so a little bit of competition won't do any harm. Some of the historic races are also pretty weak commercially so maybe even a break from the sport might give them a chance to find new investment. Not only that: if the populations of these places don't have a Grand Prix for a few years, they might miss it and take more of an interest. Some countries either forget or take for granted just how much money comes with a Grand Prix. It's huge!

One of the races that I'm sure will be kept on its toes by the introduction of Las Vegas is Monaco. In fact, in the few days since the announcement a lot of people have been saying that this new race could end up putting Monaco to shame. Monaco's biggest selling points these days are glamour and tradition. Plain and simple. Las Vegas has glamour to burn and because it's a lot bigger than Monaco it will almost certainly deliver a more exciting and competitive race. What is Monaco left with, then? Tradition. And what can tradition do if you're not careful? It can stop you from moving forward. In the old days Formula 1 needed Monaco more than Monaco needed Formula 1, but that's changed now. Being on the calendar is no longer a foregone conclusion for any country and that's how it should be. The effect I think this will have on the relationship between Formula 1 and Monaco is positive, as nobody wants the relationship to end. Then again, Monaco has been getting a lot of bad press in recent years for being boring and predictable, and the more exciting the new races are the worse this will be.

Thursday, 7 April 2022 – Albert Park Circuit, Melbourne, Australia

2 p.m.

Happy birthday to me, happy birthday to me, happy birthday dear Guenther, happy birthday to me. Today I am fifty-seven years young and I don't feel a day over ninety! People seem to be obsessed by age these days but to me it's just a number. How old I feel depends on how my team are doing, so until Bahrain I had been dead for three foking years!

Earlier today the team surprised me with a cake, which was nice of them. The only issue was that it was presented to me by Kevin, who has a disease! Are you trying to kill both of us? The cake was good, though. Strawberry. I made a speech, which sent everyone to sleep, and then did some interviews.

Just to add to the festivities, this weekend will be Haas's 125th race in Formula 1. I had to do a Q&A for our website earlier and Stuart reminded me. Perhaps fittingly then, not one but two of my most memorable moments at the team happened here in Melbourne. The two sixth places.

4 p.m.

I had to go and answer some questions on the stage in one of the fan areas. This happens at most Grands Prix these days and some of the drivers and team principals enjoy it, and some don't. I foking love it! Everybody is on the same team in that situation and are just having a laugh together. The person asking the questions told the crowd what day it was and so everybody sang happy birthday to me. There must have been ten or twenty thousand people. But not just any old people. Australians! The noise they made was incredible. It was quite a moment for me.

I've just been talking to somebody about the alterations they've made at the circuit here, which are supposed to facilitate more overtaking. Several turns are wider and they've reprofiled one or two corners. The biggest change they've made is removing the old chicane at turns 9 and 10, which has created a power-heavy section along Lakeside Drive. According to the organizers, the lap times have been reduced by about five seconds compared to 2019, despite just 28 metres being shaved off the racing line. Shit a brick!

I've just been in the garage and the relationship between Kevin and Mick seems to be growing all the time. It's nice to see. There's respect on both sides and they trust each other, which helps me out tremendously. On the flipside of this we have to make sure that they are not getting too comfortable, otherwise they won't push each other. Like everything, it's about finding a balance. If Kevin was closer to retirement age I think that might be a problem, but he's still young and competitive. Also, if he does a good job for us, one of the big teams could come in for him. That will definitely be at the back of his mind at the moment and I'm fine with that. As long as the car is competitive he will keep on pushing.

Friday, 8 April 2022 – Albert Park Circuit, Melbourne, Australia

10 a.m.

Poor old Kevin still feels like shit. I asked him if he'd like to go back to the hotel but he insisted that he wanted to drive. I don't know. Something doesn't feel right about this weekend. Kevin feels like shit. I don't feel too good. We just have to push ourselves.

4 p.m.

FP1 didn't go well at all. Kevin recorded eighteen laps with a best time of 1:23.186 (P18) but felt sick the whole time. Mick trialled a new tyre compound at the start before switching to a set of softs and delivered a time of 1:24.349 (P20). We have to do better than this.

The lap times dropped during FP1 as the track continued to rubber in. Kevin did a 1:21.191 (P16) during his qualifying simulation, while Mick managed a 1:21.974 (P18). Both drivers wrapped up the session with high-fuel runs, which gave us some valuable data for the race.

I'm starting to feel even worse now. Not nausea, like Kevin. I just feel dizzy and have a headache. I thought at first it might have been something to do with the jetlag (that I'm not supposed to have!) but I'm pretty sure it's not. This is all I need. I think I'll go straight back to the hotel.

If this book ends on this page it's because I died during the night. I just thought I'd warn you.

Saturday, 9 April 2022 – Albert Park Circuit, Melbourne, Australia

7 a.m.

Well, I didn't die during the night. That's about the only positive I can find, though, at the moment. I didn't sleep well and I still feel like shit. I'm never ill. I'm like that foking rabbit from the battery adverts. Just with longer legs and a bigger nose. Anyway, I'd better drag myself out of bed and see what happens.

4 p.m.

Jeezoz. It's safe to say that qualifying didn't go according to plan. Kevin managed to work his way from a 1:21.243 to a 1:20.548 on his first set of tyres and managed 1:20.254 on the second set, putting him P17. The third and final run was a sprint to the line after a foking red flag but Kevin was unable to improve his time, which put him out in Q1. First time this season. *Merda!*

Mick fared slightly better, fortunately, and advanced to Q2 courtesy of a quick lap of 1:20.104 (P15). Good start. I thought he might have been able to improve on that but at the end of Q2 he was still P15.

I now have a load of media to do. I wish I could just go to bed.

Sunday, 10 April 2022 – Albert Park Circuit, Melbourne, Australia

6 p.m.

Mick finished thirteenth and Kevin fourteenth. Set-up wasn't as good. Race pace was OK but the safety car foked things up for us. It happens. Let's just say that today has been a learning curve. No idea how I managed to stay on my feet. Could happily have crawled under my desk and gone to sleep for a couple of weeks.

Monday, 18 April 2022 – Steiner Ranch, North Carolina, USA

2 p.m.

Well, have I got a story for you guys. What a couple of weeks it's been. Not in a good way. In a really shit way!

After qualifying on Saturday I felt slightly better so I went out for dinner with Mattia. We had a little bit of wine but nothing heavy. The next morning I woke up and I felt like I'd had a bottle of whisky! Seriously, I felt so drunk. When I went to the track I went straight to my office and as soon as I sat down I started dozing off in my foking chair like an old man. Me? In all my years in motorsport, that has never happened before. I'm always ready!

After the race I went straight to the airport and in the car on the way there I fell fast asleep. It took the driver five minutes to wake me up. He thought I was dead! When I got on the plane I fell fast asleep again and didn't wake up until we landed in Los Angeles almost ten hours later. After getting off the plane I went to the lounge to wait for my connection to Charlotte and fell asleep again until my flight was called three hours later. When the person at the lounge woke me up, I said, 'Oh shit, I hope I wasn't sitting there with my mouth wide open all this time.' 'Actually, that's exactly what you did, sir,' she said, a little bit too honestly. By this point I was shaking so much that I could barely hold my cell phone and when I got on the plane to Charlotte at a quarter to midnight I fell fast asleep again. It might have been a red-eye flight for some people but not for me. It took the cabin crew ten minutes to wake me up when we were landing in Charlotte. They were really worried, apparently. I knew nothing about it because I was away with the foking fairies!

I eventually arrived home at 7.30 a.m. and I decided to have a bit of breakfast. I didn't feel too bad at the time but still felt tired so I had a shower and went to bed. The next thing I remember is Gertie waking me up. 'You know what the time is, don't you?' she said. 'It's one p.m.' 'I don't care,' I said. 'Just let me sleep.' I think I slept for another five hours, so by the time I actually came back to the land of the living again I'd been asleep, on and off, for almost twenty-four hours.

On Tuesday I woke up and was sick almost straight away. I only just made it to the bathroom. The nausea carried on pretty much for the rest of the day and by the evening it felt like my head was about to explode. It was terrible. I'd never felt as bad in my life. On Wednesday it was pretty much the same and on Thursday and Friday I started to feel better. Not enough that I could do any work, but enough that I could walk and talk. Then on Saturday I opened up my laptop for a couple of hours and yesterday I did a full day and started catching up on the bullshit. Good thing is that because it's Easter things are pretty quiet at the moment.

Do you know how bad it got, though? I even ended up putting an out of office message on my email. I'm not sure how long I've been sending emails but I've never had to do that before. I did go and see my doctor last Wednesday and he said it was a viral infection of some kind. He also said that if I didn't feel better by Friday I should go to hospital, but fortunately I did.

Anyway, the good news is that Guenther's back. Fitter, stronger and prettier than ever. Later today I'm flying to Italy and staying at my mum's place for a night, then I'll drive down to Imola for the next race. The troble is that this is one of about three Grands Prix that I really don't think will suit our car, so to be honest with you I'm not expecting great things. Then again, at the Grands Prix where I expect us to do well we've sometimes done shit so using that logic we'll probably foking win! We've got the first sprint race to look forward to this weekend. That will also be a challenge because we've only got one Free Practice session to find a good set-up. But it's the same for everybody and so the people who are best prepared will have a better car. We'll be ready, though. Bring it on!

Thursday, 21 April 2022 – Imola, Bologna, Italy

2 p.m.

There's been more talk this week in the press about our relationship with Ferrari than there has been about the Grand Prix. Seriously, this is getting boring now. I think they overestimate my influence at Ferrari, not to mention the access I have at Maranello. What do they think I'm doing there, sneaking into the Ferrari offices at night and tapping into their servers? We are in a completely separate building to Ferrari at Maranello. How many times do I have to say this? Mattia has come out and made a statement but I don't think it will do any good. He said that Haas is a fully independent team compared to Ferrari and, as well as us not being a junior team, they are not exchanging any information beyond what is allowed by the regulations. That's pretty to the point, I think.

We can't win, though. When we have a good car, it's copied from

Ferrari, and if we have a bad car, that's where we belong. I don't think this way of thinking shows intelligence. You also have to respect people and that is lacking in certain places. We are not here to fill up the grid. If someone has done a good job, you should take your hat off to them and say, 'Well done.'

One of the most vocal opponents of our relationship with Ferrari has been Andreas Seidl from McLaren. He believes that manufacturers should only be allowed to share power units and gearboxes with other teams and that teams like Haas should only be allowed to exist if they do everything else in-house. What he wants, then, is basically a return to the good old days when you had lots of teams overstretching themselves and going to the foking wall. Either that or he just wants four or five big teams.

All I can do is remind everybody like Mattia did that everything we are doing is within the regulations. And if Andreas has a problem with that, he should moan to the FIA about it. In fact, I have formally invited the FIA to come and investigate the relationship between Ferrari and Haas, and if they find anything wrong then we'll put it right. They won't, though, as everything we are doing – yet again – is within their own regulations.

I found out a few weeks ago that Andreas was the person who had argued against us receiving extra testing time in Bahrain, even though it wasn't our fault that the freight was late arriving. When I found out I thanked him for being a sportsman and asked him why he would do something like that. He just laughed at me. I then told him exactly what would happen if he laughed at me again and he stopped. I'm fine with people having different views to me about the way teams are set up and operate, but don't you disrespect me or my team. I won't stand for it.

Friday, 22 April 2022 – Imola, Bologna, Italy

4 p.m.

After all the shit yesterday surrounding us and Ferrari, not to mention what happened in Australia and then me turning into Sleeping Beauty for a foking week, I really needed something good to happen today.

And it did!

In qualifying for the sprint race tomorrow we had our best ever result when Kevin secured P4. P-foking-4, baby! At one point I thought it was going to end in disaster when Kevin spun off the track, but he was able to keep the car going and qualified behind Verstappen, Leclerc and Norris. There was some serious noise coming from our garage after that, let me tell you. Not to mention the crowd. By the way, did you know that Leclerc is a former Haas boy? It's true. He was a development driver for us in 2016 when he was driving in Formula 2 and made a couple of appearances for us in FP1. I like Charles. He's a good guy.

Just like Bahrain, everyone is smiling again. Also, as well as making all the right people happy, it will also piss the right people off. Sorry, I shouldn't get political. I'm still a bit angry about that.

Tonight Stefano Domenicali, who is from Imola, is taking all the team principals to his favourite restaurant here. As you know, Stefano is a good friend of mine and he'll make sure all of us have a good time. I bet he's careful about the seating plan, though. He's not daft. Christian will be at the opposite end to Toto and I will be at the opposite end to Andreas. Or, at least, I hope I will be. I don't want to spend all night arm wrestling him and kicking him under the table. I actually quite like Andreas. He's an OK guy.

A lot of people ask me if the whole Christian versus Toto thing that you see on *Drive to Survive* is completely for real. 'Do they really hate each other?' Blah blah blah . . . If those guys aren't cuddling all the time, why should I care? I said earlier that the thing that stops us from killing each other is that, despite us working for different teams, we are all working for, and all care about, Formula 1. Toto and Christian are a case in point. Sometimes in the moment I think they probably do want to kill each other, the same as I would sometimes like to kill the little shit who questions our relationship with Ferrari. At the end of the day, though, we are all in this together and as much as I disagree with what this guy thinks, just like Christian will disagree with Toto sometimes and vice versa, we have to remember that ultimately we all sit under the same tree. If we didn't remember that, we'd be kicking the shit out of each other all the time.

Actually, that's not a bad idea. If we are ever having a boring race we could organize a white-collar boxing match between two team principals. Could you imagine that? Poor old Fred, though. He couldn't knock the skin off a foking potato. He's actually a couple of years younger than me but looks like he could be my grandfather. I'm not sure who would be the toughest. I've known Christian a long time and used to work with him at Red Bull. He could be pretty tough. Then again, he's married to a Spice Girl now. Otmar at Alpine looks like he could be pretty useful also. He's a big guy.

Anyway, back to this evening. I think if Stefano puts me next to Fred there won't be any blood. Just a lot of taking the piss. The food and wine will be good. And it's a free meal. What's not to like?

Saturday, 23 April 2022 – Team Hotel, Bologna, Italy

7 a.m.

I'm afraid last night went pretty much without a hitch so there is nothing to report. I know, I'm disappointed too. Maybe next time. It was good fun, though, and the food was unbelievable. Stefano was also a fantastic host.

Obviously we have high hopes for today. Yesterday's performance by Kevin was one of the best he's ever given us in a Haas car so we'll be expecting more of the same today. Mick qualified twelfth, by the way, which wasn't too bad. He's still yet to make a Q3 though, which I know he's desperate for. It will come.

Anyway, now I'm going to have a shower, put my lucky shorts on and go straight to the track.

4 p.m.

I'm not in the best of moods at the moment so I'm going to keep it short. All that promise from yesterday came to very little in the sprint race. Kevin managed to stay in fourth until lap eight. Three laps later he lost fifth to Ricciardo. By lap sixteen Kevin was in seventh and Mick in tenth, so we had one driver still in the points and one driver just outside. Kevin then dropped one more place before the end and Mick stayed in tenth, so that was one point in total. Some people might say, 'Come on, Guenther, a point is a point.' Yes, but when you qualify fourth? It's not a disaster but I expected more. The one positive is that Mick will start tenth tomorrow, which is encouraging.

Anyway, I'm going to bang my head against a wall. Fingers crossed for tomorrow.

Sunday, 24 April 2022 – Imola, Bologna, Italy

6 p.m.

So near, and yet so far away – again! Kevin had a great start and moved up to fifth but then Mick made contact with Alonso in the first chicane, went into a spin and came out of it seventeenth. Kevin tried really hard but the fact of the matter is he wasn't as fast as the other guys. At least race pace is now the issue and not reliability! Kevin finished ninth in the end, which gave us two more points, and Mick stayed in seventeenth. I'm a bit disappointed in Mick. That was a great opportunity for him. It's just down to experience really. A lot of people are starting to say that the car is better than the results we are getting and at the end of the day that is down to the drivers. I'll give it a few more races but something needs to happen. I'd like to talk to Mick now but he's flown straight to America. Am I thinking about alternatives already? Of course I am. Kevin has scored fifteen points so far this season, and if Mick had scored even half of that we'd be on twenty-three. We've waited a long time for a good car and the only thing that can really spoil the party is if we don't realize its full potential. Nothing would be more frustrating than that.

OK, it's Miami next. A new race. I wish I was going home first but tomorrow I fly to London for some meetings and then I'll be at the Oxford factory for a few days. The way I feel right at this moment is quite new for me at Haas. Happiness or disappointment are the two normal emotions in my life but I'm switching from happy, to frustrated, to nervous, to disappointed and then to happy again. It's a strange combination, and with Miami being new I can't see that changing anytime soon. It sounds like I'm complaining but I'm not. The frustration means that we're close to something good. And we are. I can feel it!

Monday, 2 May 2022 – Haas F1 Factory, Banbury, Oxfordshire, UK

A bit of good news. Tomorrow we will be announcing a new partnership with a financial technology company based in Denmark called Lunar. We've been working on it for a while now and it's just about ready to go. Although we're in quite a good place financially, the Uralkali deal was worth a lot of money and I've been working hard to replace it. About 65 per cent of our annual budget is normally covered by sponsorship income so you know the kind of figures we're talking. It's big money. It's early days but there are already a good number of interested parties for the title sponsorship. It's not completely essential but an American company would be nice.

I think some people in marketing who aren't Formula 1 fans assume that all the teams are the same, or are at least variations on a theme. I always like to correct them on that. 'We're a little bit different to the other teams,' I say. And I'm not talking about money. We have a different ethos to the others. Let me give you an example. Every other team on the F1 grid apart from Haas is

corporate. Or at least to a certain degree. Mercedes are totally foking corporate, which is what you'd expect from an uber-powerful
German outfit that has lots of money and has a guy like Toto in
charge. They're the real deal, those guys. The Alfa Romeos and
the Alpines of this world are less corporate, but it's still part of
their DNA. Haas, on the other hand, are not corporate. We're not
anti-corporate. Of course we're not. It just isn't what we're about.
A few weeks ago a journalist asked me what our USP is.

'What separates you from the other teams?' he said. 'What makes
you different?'

'That's a good question,' I replied.

I then mentioned the fact that we're not really corporate and at
first the journalist didn't get it. 'But F1 is the most corporate sport
in the world,' he said. 'Surely you have to follow that lead.'

'Bullshit,' I said. 'An F1 team can be whatever you want it to be
as long as you stick to the rules, and the rules do not state that you
have to be corporate.'

It isn't for everybody, though. If you go to Mercedes or Ferrari
as a guest there'll be an army of people to look after you. Whereas
if you come to Haas as a guest, we'll pass you some overalls or get
you to wash up! Not really. Not yet, anyway. Let's see how the new
season goes and how many foking chassis the drivers go through.
The fact is, though, that we don't have lots of people to look after
our guests so with Haas it's a different experience.

What else makes Haas different? Well, some people would probably suggest that I am different to the other team principals. And
they'd be right. Not in a bad or good way. Worse or better. I'm just
different. Can you imagine me at Mercedes? Jeezoz Christ! The
mechanics there start collapsing just when I walk past the garage.

The whole of Germany would spontaneously combust if I went there.

The main difference about Haas isn't me, though. Or the fact that we're not corporate. The difference is all of us. A team is only as good as the people it has working for it and we've got some of the best. Not everybody is right for Haas, though, and Haas isn't right for everybody. That's why I make sure I'm involved with all the hiring and firing to some degree. I usually come in at the second interview stage. Everybody, no matter who they are, is answerable to me, so I have to make sure that they're the right fit. I don't always get it right, but it helps that I'm involved. Or at least I think it does! The other teams have the corporate gene, and we have Gene Haas. A quiet man who loves his motorsport and was brave enough (and mad enough) to put his trust in an ugly, loud-mouthed Italian who would probably foking die for the team that he helped to create.

You have no idea how many naysayers there were when we were starting up. 'You'll be out of the sport quicker than it took you to get in,' they all said. Bunch of wankers. They also assumed that because we were the smallest team it should be something to be embarrassed about somehow. I'm not embarrassed to be the smallest team. I'm proud about that. If you can do more with less, that's a good thing, don't you think? It's certainly more of an achievement. We're not poor, though. We just try and manage our money properly because we want to be here for a long time. All good things come to an end, but until that happens we want to make the best of it.

The new rules have certainly helped, partly because they've been designed to encourage the teams to operate as profitable enterprises as opposed to money pits or a rich man's hobby. Formula 1 has been crying out for this, but of course it presents a challenge. Any fool can run a business at a loss.

Tomorrow I fly to Miami for the first ever Miami Grand Prix. Because it's another home Grand Prix for Haas, the guys in our creative department have designed a poster for the race that is based on the cover of a Grand Theft Auto game called 'Vice City'. I've heard of Grand Theft Auto but I've never played it. I've got better things to do with my time, like running a foking F1 team! I'll say this, though – ten out of ten for creativity. It looks incredible and according to Stuart it's gone viral online, whatever that means. I am wearing a pair of sunglasses on it and look pretty cool, to be honest.

So, how about a bit of background on the race? Miami was first suggested by Stefano as a possible venue for a Grand Prix at the beginning of 2021. Based on the number of people who were already stopping me to talk about the sport and ask for selfies in places like Walmart, I already knew that the audience was there. Miami and F1, though? That's a perfect combination in my opinion and I was really excited.

When Austin was first announced seven or eight years ago, a lot of people said that it would be lucky to last two years. Even I worried that it might not become a regular fixture. And look at it now. Austin has already become a classic and we get the same number of people to that race now as we do to Silverstone. Four hundred thousand people over a weekend. Wow! That's just incredible. *Drive to Survive* has to be given credit for generating a lot of the initial interest in F1 in the States. For sure, that's the case. But if the people did not like the sport they wouldn't come back, and they are coming back. Every year there are more.

Formula 1 is one of the very few major sports, apart from soccer, that has true global appeal, but until recently that hasn't been realized very much. Before it was just petrol heads who followed F1, whereas nowadays everybody you speak to seems to have a favourite driver or team. Has it ever been in a better position? No, I don't think so. There will be some purists out there who think differently to me but in terms of popularity and excitement I think Formula 1 is in a good place at the moment. The only thing it has to be careful about is ensuring that the sport itself remains paramount and not the show. Everybody likes a spectacle but if it's compensating for something boring or shit then people will eventually look elsewhere and move on.

Do you know what I found out the other day? The hotels in Las Vegas are sold out for the Grand Prix there next year. Over a year and a half in advance! There are a hundred and fifty thousand hotel rooms in Las Vegas and they're gone. All of them. I asked our travel coordinator, Kate, where the team were going to stay next year and she said to me, 'At the moment, Guenther, I'm not sure!' It's foking crazy. I could be in a caravan in the desert at this rate.

Anyway, I better get my shit together as my car will be here soon. See you in the Magic City!

Wednesday, 4 May 2022 – Miami International Autodrome, Miami, USA

6 p.m.

I arrived in Miami at 3.30 p.m. and went straight to the track. Everything about the set-up here has surpassed my expectations.

It's foking amazing. The circuit is temporary like Albert Park but feels like it's permanent and is set in the Hard Rock Stadium complex in Miami Gardens. It's where the Miami Dolphins play, and if you go to the very top of the stadium you can see every corner of the track. It's very impressive.

There's already been talk about whether three Grands Prix will be enough for the American market and it's something I've already been asked about quite a bit since the Miami Grand Prix was announced. My opinion is that they need to consolidate what they have first for a few years and then look again. One of the three races hasn't even taken place yet. Just calm down!

Being here, talking to the public and seeing what they've achieved here demonstrates the change that Formula 1 is going through at the moment. When Haas first started, if you told somebody in the States that you worked in Formula 1 they'd look at you and say something like, 'That's just a British sport, isn't it?' Nobody cared about it. Now when you tell them, they're like, 'Wow, Formula 1. That's really cool.'

From the moment I got off the plane I could tell that this race is to be something special. I arrived on Tuesday and already there were fans everywhere looking at the circuit and creating an atmosphere. I spoke to one of the TV guys earlier from ABC and he thinks that the viewing figures from the Miami Grand Prix could even rival some Nascar races. If that happens I think my head will foking fall off. It would be amazing. Last year Austin got the second-highest USGP viewing figures ever in the States, with 1.2 million. As impressive as that is, the average for Nascar is about 3 million a race so there's a way to go. Miami is new, though, and the curiosity value will be massive.

Friday, 6 May 2022 – Miami International Autodrome, Miami, USA

8 a.m.

This week, Ralph Schumacher, Mick's uncle, has claimed that I have a better relationship with Kevin than I do with Mick. And he's right, for once. One thing he doesn't do, though, is ask why I have a better relationship with Kevin or even ask if it's relevant. So let me tell you. For a start, I have known Kevin for a very long time and we've been through a lot together. I know how he works, he knows how I work and we have an understanding. Or at least, as much as you can in such a fluid and high-pressure environment. It's not that I have a bad relationship with Mick. Far from it. It's just not a very extensive one. I have tried to get to know him a bit better but at the end of the day it takes two to tango. You can't force it. Maybe it will happen over time. Comments like Mick's uncle's don't help, though, and my relationship with Mick has no bearing on how either of us is performing. If we were arguing all the time or couldn't be honest with each other then it might, but that isn't how things are. Just because I'm not close to Mick doesn't mean I can't be frank with him, and vice versa. Another day, another headline.

Anyway, I'm going to the track. This is where the fun begins.

6 p.m.

The track temperature during FP1 was 54 degrees and the air temperature 34. That's pretty hot! Mind you, we're in Miami. What do you expect?

Both drivers did twenty laps and then in FP2 about the same again. Kevin finished tenth overall and Mick fifteenth. It was an interesting session and we've learned a lot. I think we can still improve but, all in all, I'm pretty happy with what we gained today in knowledge. I don't think we're in a bad place so, again in FP3, we'll make some progress and get ready for qualifying.

There's been a lot of talk in the paddock today about which celebrities are going to be turning up over the weekend. I don't really watch television or listen to music much so the chances of an old fart like me recognizing anybody are tiny. In 2019 this guy called Trevor Noah was a guest of ours at the USGP and I had no idea who he was. 'He's a South African comedian,' said Stuart. 'Really,' I said. 'I've never heard of him.' 'Well, he's got 12 million followers on Twitter and 8 million on Instagram, so he's pretty famous.' 'No shit?' I said. He was a really nice guy. That happens a lot these days. 'Guenther, this is blah blah.' I know they're famous but I have no foking idea who they are half the time. I live in a bubble, you see, so unless you're famous in motorsport, I won't have a clue who you are.

Saturday, 7 May 2022 – Miami International Autodrome, Miami, USA

1 p.m.

I was right about the celebrity VIP thing. Apparently the paddock is full of them today but I haven't recognized one yet. We've actually got a few of our own coming this weekend. The Watt brothers, JJ and TJ, have been guests of ours in Miami. They're very famous NFL players and are really nice guys. They're also foking enormous! This is already like Monaco times ten. It's off the scale. God knows what Las Vegas will be like next year. And I get it, you know. Celebrities want to be associated with something popular and glamorous, and Formula 1 wants to be associated with popular and glamorous people. Let's face it, it needs some help in that department. Look at the ugly bastards we have at the moment.

What I don't agree with is celebrities not wanting to talk to any of the media people. I don't like that. They're all right walking up and down the grid having their photos taken and being made to feel important, but the moment a press or media person says, 'Hi, are you enjoying yourself today?' the bouncers rush in and start losing their shit. In my opinion, those people should be made aware of who the main press and media guys are in F1 and they should be encouraged to chat to them. If you're here on a free day out you should support F1 by interacting with the people whose job it is to discuss and promote it. That's the deal. Or at least, that should be the deal. Our house, our rules.

That celebrity self-importance thing has only appeared over the past few years and I'm really hoping it will go soon. I've seen some of the shit that guys like Martin Brundle have to put up with on the grid and it's not good. I see him sometimes getting trampled on by bouncers. It's sad. Our sport should be for everybody and when the only people who don't want to interact are not even part of the sport, that isn't a good look. This is a Grand Prix, not the foking Oscars.

6 p.m.

Not a good afternoon. In fact, it's been a pretty shit afternoon. In FP3 we took one step forward by putting in a strong performance and then we took two steps back courtesy of a malfunction and a couple of bad decisions. At the start of Q1, Kevin's radio malfunctioned, which meant he couldn't communicate with his engineer. As a result, he remained on-track on a single set of soft tyres and finished P16. Mick, who was on the same strategy as Kevin, managed to come in and change his tyres and scraped through to Q2. He couldn't really improve, though, and ended up qualifying last in P15.

We still have no idea why Kevin's radio stopped working. It's a mystery. An infuriating mystery! Had it been working he'd have been able to change tyres and the chances are he'd have made it to at least Q2. Anyway, shit happens. To us. A lot!

Sunday, 8 May 2022 – Miami International Autodrome, Miami, USA

8 a.m.

Last night after qualifying I got the team together for a talk. I said that we have an opportunity here and we need to step it up a gear. It feels like we're hanging back at the moment. Like we're banking on tomorrow being a better day all the time. Well, as I said to the guys, those days are going to run out pretty soon and unless we pull our fingers out and start turning things around, all that promise will evaporate and the season will turn to shit. We need to work harder and, more importantly, we need to work smarter. People sometimes get a little bit down when I have one of my rants but I made it clear to everyone that the reason I am saying this is because the success is in touching distance. They, just like me, would be devastated if we got to the end of the season knowing we had a good car but didn't manage to realize its potential.

After that meeting I got the engineers and the drivers together. I said: 'Do you understand what I mean, guys, because if you don't we may as well go home now. Seriously, you cannot put your head in the sand and wait for things to get better. We have to make them get better together and we have to do it now. If I've said it once I've said it a thousand times. We are not just here to make up the numbers.'

A theory as to why this might be happening, which is something else I discussed with the drivers and engineers, is that we could still be suffering the effects of last season. Last year everything was negative. Everything. There were no positives to be had. I tried reassuring them that better times were ahead but as much as they might have wanted to believe that, the only thing that was real and guaranteed was the fact that next week would bring around another disappointment. Now that isn't the case any more it's almost as if they are scared of being competitive again. They're scared of the car and they're scared of themselves. I think the guys could see what I was getting at here and realize that it's a straightforward confidence issue. Just as we got used to expecting to finish last, we now have to get used to expecting to finish in the points. That has to be the mentality going forward. We need to be hungry.

'Come on, guys, it's in our own hands!' That's basically what I said to them.

I'm quite nervous about today, I'll be honest with you. Things need to turn around quickly.

10 p.m.

I'm afraid I couldn't write this earlier as I was in what you might call a bit of a bad mood. Actually, that's an understatement. I was spitting foking feathers!

Where do I start? It was actually quite an exciting race in parts. Or at least until the safety car came out. Both drivers made up places after the start and it was encouraging. Then, after the safety car, Kevin wanted to go on new tyres. The race engineer and strategist advised him not to because the advantage wouldn't be big enough and they thought he could end up losing places. Kevin was adamant, though, and in the end we decided to relent. As feared, he ended up losing places after all and it probably cost us points. To be fair to Kevin, he held his hands up straight afterwards and apologized. It didn't make it right but at least he might have learned something. Mick ended up getting into an incident with Vettel after the safety car, which ruined his chances, so all in all a shit afternoon really. At one point we were lying eighth and ninth and really we should have stayed there.

Immediately after the race it felt like everything I'd said to them the previous evening had been ignored. That obviously isn't the case, though, and fortunately I realized that before giving them an encore! I was just feeling a bit pissed off so went back to the hotel and put my head down the toilet.

Monday, 9 May 2022 – Charlotte Motor Speedway, North Carolina, USA

10 a.m.

The things I do for this team!

I don't know whose idea it was but while we were gearing up for the Miami Grand Prix, somebody at Haas F1 suggested that Kevin and Mick pay a visit to the Stewart-Haas Racing factory one day to say hi. Partly as a bit of a PR exercise, but also because it might be interesting for the guys. Stewart-Haas are basically our neighbours and, although we're completely separate entities, we obviously share an owner. Gene has been in Nascar since 2002 and he set up Stewart-Haas Racing with Tony Stewart in 2008. Their factory in Kannapolis, which isn't far from our own headquarters, measures 140,000 square feet and is very impressive. It's a great facility. Kevin and Mick went to the factory and said they had a great time.

While this was being set up it was also suggested that we use the day for Kevin and Mick to take a Nascar out for a spin at the nearby Charlotte Motor Speedway. Kevin and Romain had done it back in 2019 so we said, 'Why not?' Kevin was keen to do it again and Mick and Pietro wanted a go, too. One of the Stewart-Haas drivers, Chase Briscoe, was also going to be there. It sounded like fun. 'You're coming too, though, Guenther?' 'Sure,' I said. 'I wouldn't miss it for the world.'

When we arrived here about an hour ago the person from Stewart-Haas who had organized it all said, 'I'll tell you what. Why doesn't Guenther have a go?'

'You what?' I said. 'I can't drive one of those things!'

The closest I've ever been to racing cars was when I was a co-driver for a rally team in the early 1990s. But only for a few races. That's as far as it ever got. I never had the talent or the money to become a driver.

'But I drive a Toyota Tundra, for fok's sake!' I said.

'How about Kevin takes you out, then?'

Had this been on the back of us scoring some points in Miami then I would have been all for it, but it wasn't.

'The last time I spoke to Kevin was after the Miami Grand Prix, and he'd just had to retire after colliding with two other drivers. It was quite a frank conversation and also very honest. What if he wants revenge?'

And that's basically where we are now. I'm sitting here at the Charlotte Motor Speedway waiting for the Dane with a vendetta to come and take me for a spin at 200 mph. What could possibly go foking wrong?

Shit, that was cool. *Really* cool. I don't think I can turn my head to the right, though. No, I can't. Jeezoz! It almost didn't happen as I couldn't get in the foking car. I forgot that the doors are welded shut so you have to get in through the window. The best part was going around the oval. I've never experienced that before and I'd do it again in a heartbeat. To be brutally honest with you, I do have a little bit of experience driving quickly with other people at the wheel, as I went out quite a few times with the legendary Colin McRae. That was a while ago, though, and I haven't done it since. As much as I respect the guys who are here today, Colin could do things with a car that would make your foking eyes bleed! That guy was a genius in my opinion, and after Monaco I will try and remember to tell you a story or two from those days.

I got the viewing figures through for Miami earlier and they're impressive for sure: 2.5 million in the States. That's creeping up to Nascar figures. I think it could be at least that much for Austin in October and then God knows where we'll be next year with Las Vegas on the calendar. These are exciting times for Formula 1.

Thursday, 19 May 2022 – Team Hotel, Barcelona, Spain

6 p.m.

Tomorrow it will be three years since we lost Niki Lauda. I can still hear his voice. 'Guenther, it's Niki. OK, listen to me!' He is the person who first persuaded me to leave rally and come and work in Formula 1 and so it seems like a good opportunity to tell you the story of how it happened and then what happened next. That's not quite true, actually. What I should have said just then is that he was the person who *told* me that I was going to leave rally and come and work in Formula 1, and work for him.

When Niki took over Jaguar in 2001, he had a big job to do and was looking for new people. I was working for M-Sport at the time, who had been commissioned by Ford, who owned Jaguar, to work on the new Focus for their works team. The motorsport manager of the works team, a guy called Tyrone Johnson, gave Niki my name and one day I got a call from Niki's secretary saying that Mr Lauda wanted to speak with me. I had never before met Mr Lauda but I had been a fan of his as a child and so it was an honour just to be asked even to speak to him.

Our first conversation was over the telephone and, after asking me a few questions, he said, 'I don't suppose you are going to be in Vienna any time soon?' Fortunately we were about to compete with the Focus in the Austrian Championship so I said that I could meet him in a couple of weeks. 'We will have dinner,' he said, and then he put down the phone. There were no hellos and goodbyes with Niki. When he was finished, you knew about it.

I'd been told by several people that Niki didn't do long dinners and so was expecting it to be over within half an hour. In the end we were there for over two hours and at eight o'clock the next morning he called me up. 'Thank you for your time yesterday,' he said. 'You will be working for me.' 'May I ask you what I will be doing?' I replied. 'I haven't worked that one out yet,' he said. 'I'll let you know.' And then he hung up.

I still had a job to do with the Ford works team so didn't get to join Niki until the following January. When I did join, I did a variety of different jobs, and then one day I got another call. 'Guenther, you will now come and run the team with me.' 'Really?' 'Yes.' End of conversation.

I could probably write a foking book about what happened during my time at Jaguar. It was a bit of a disaster, though. A real mess. To cut a long story short, a big corporate company like Ford taking over a race team wasn't a good idea because they thought it was going to be easy. There was no real understanding. By the time Niki arrived, a lot of the damage had already been done and Niki's vision for the future didn't match with Ford's. He was an old-fashioned racer and Ford were very corporate. It wasn't a good match.

By the time they decided to let Niki go, we had started to make some progress. I felt loyal to him, though, so it was decided that when he went, I would go too. Ford then started to lose interest in the project and in 2004 it was bought by Red Bull.

I hope I'm not boring you here? I know this is supposed to be a diary but I get asked about this a lot and I am thinking that, if everybody reads it, it might save me some time in the future.

By the time Red Bull bought Jaguar, I was working for the Opel DTM team – as a contractor, though, so I wasn't tied to them. This meant that when I received a telephone call from Mr Mateschitz offering me a job with his new team, I was able to say yes. I had no real ambition to go back into Formula 1, though. I was perfectly happy in DTM, to be honest.

I was still living in the UK when all this happened, but because the new team were situated in Austria it made sense for Gertie and me to go and live near my family in South Tyrol.

When I started at Red Bull I had no idea what to expect, really. I didn't know what Red Bull were like as a company and my only fear, I suppose, was that perhaps they too were very corporate. In the end, not a lot had changed since Niki and I had left, which was a surprise. Certainly when it came to the structure of the team. What had changed, and what made the difference, was the level of investment that Mr Mateschitz was putting in.

For the next couple of years I ran the team with Christian Horner and the results improved. Then, in late 2005, Mr Mateschitz asked me if I would be interested in moving to the United States to set up a Nascar team for Red Bull. That was a foking surprise! I remember speaking to Gertie about it and she was pretty keen. Also, it had always been a bit of an ambition to move to the United States, but I never thought anything about it as it was so difficult for European people to move there. This, then, was the opportunity that I never thought would happen. The only slightly daunting aspect was having to start from scratch and build a new team from the bottom up. And in a sport that I was not really familiar with. That's not easy in any country, let alone in a country that you know very little about apart from what you've seen at the movies. It was a challenge, let's say.

After arriving in the United States, it didn't take me long to realize that my experience working in Formula 1 meant very little in this situation. At least initially. Formula 1 is the pinnacle, so you think you know it all, but actually I knew shit. Realizing this early on probably saved me — instead of ploughing ahead and making a mess of everything, I just sat back and observed for six months. I was like a foking sponge and took it all in. The culture, the way they do business. Even the way people speak to each other day to day. I wanted to know everything.

After about six months I was able to start introducing little improvements based on what I had learned in Formula 1. For instance, the Red Bull Nascar team was the first of its kind to have a full-time pit crew. And they were not mechanics. They were athletes who only did pit stops. I also introduced an engineering structure, which hadn't been done before. Now every Nascar team has one in place.

When I started to implement these things, I experienced quite a bit of resistance. A lot of people thought that Nascar wasn't ready for these changes but in my opinion they were fundamental to the sport moving forward.

A common question when I am talking about my time in Nascar is whether I thought that I would ever come back to Formula 1. I don't really think like that. If an opportunity comes along for me and I like the look of it, that will be my focus. I don't look backwards and I don't look that far forwards. I still had lots of friends in Formula 1 when I moved to the United States and I still had a lot of friends in rally. But I didn't stay close to those friends because I thought I might go back again one day. They were just people I got on with.

The job with the Red Bull Nascar team only lasted a couple of years in the end. I don't want to go into the details about why I left but let us just say that it had run its course. I soon got some offers to move back to Europe and, although they were very attractive offers, I had a feeling that we were not done with America yet. Or that America was not done with us. In two years we had built a good life there. Gertie was happy and so was I.

OK, this part will really send you to foking sleep. Don't worry, I am almost finished. After leaving Red Bull I decided to fulfil a dream by starting my own company. I'd got to know the American motor racing scene pretty well by then and had noticed that the composites industry was potentially underdeveloped there. I was comparing it to the European market and it was like chalk and cheese. To cut yet another long story short, with the help of a friend of mine called Joe Hoffman, who had been my manufacturing manager at the Red Bull Nascar team, we started a composites factory called Fibreworks, which we still own and which is doing very well. We launched the company in 2009 and for the first five years I worked very hard with Joe getting it off the ground. Then, as you know, I started putting together the idea of starting a Formula 1 team, and the rest is history.

These days Joe runs the company while I travel around the world swearing at people and having my photograph taken. I am very good at setting a company up and all of the initial stuff, whereas Joe is very good at management and all of the day-to-day things. In fact, he is brilliant at it and the arrangement that we have works very well. Fok it, while I am here I will put in a little advert for the company. That will please Joe.

Fibreworks – leader in the composites industry. On the cutting edge of custom design, engineering and manufacturing.

Putting together a Nascar team and then a company in America definitely helped when it came to setting up a Formula 1 team there. You have to understand the culture. They might speak the same language as the British, for example (well, kind of), but they are like chalk and foking cheese when it comes to how they do business. It's not bad or worse than Britain. Just different. So, when it came to seeking investment I was able to say the right things. For a change! If I had not had that experience there is no way it would have happened like that.

So, there you go. There are a few minutes of your life that you will never get back. Even though I said that I had no intention of coming back into Formula 1, I'm very glad I did. I am also very grateful to Niki, who was an incredible mentor and a very good friend.

The last time I spoke with Niki was 21 February 2019, two months before he died. It was the day before his birthday and I figured that it would be easier to get hold of him on that day than on the day itself. I knew that he hadn't been doing well and the first time I tried him he didn't answer. One of his kids, who I get on with very well, must have seen a missed call from me and they called me back and put him on to me. He sounded very fatigued but at least I got to say happy birthday.

Working for Niki Lauda was an experience. He was an extremely hard taskmaster but you could gain his trust simply by working hard and always doing your best. Once you had his trust, he would back you all the way. Having the support of somebody like Niki Lauda made you feel like you could do anything. He was inspirational.

Before I go, I'll tell you a funny little story about Niki. When I went to see the film *Rush*, about him and James Hunt, I spoke to him afterwards to tell him how much I enjoyed it.

'The actor who played you, Niki. He was foking brilliant!'

'I trained him,' he said defiantly.

'What, to act?'

'In this case, yes. He spent three days with me.'

'And in that time you taught him to act?'

'I taught him how to be Niki Lauda.'

'In three days? You're bullshitting me.'

'No. The reason he is so good is because I trained him to be me. That is why.'

'Nothing to do with being a good actor, then?'

'No. It is my work.'

'OK, then. Congratulations!'

'Thank you, Guenther. I'm glad you enjoyed my film.'

Friday, 20 May 2022 – Circuit de Barcelona-Catalunya, Barcelona, Spain

4 p.m.

Something that occurred to me this morning when I arrived at the track was the fact that every race has a different personality these days. This is intentional on the part of F1 and it's a great idea. Seven or eight years ago, when we started, everything had become a bit homogenized. I'm not sure whether that was also intentional but there was very little to differentiate between the races. And you know what they say – if there are always the same biscuits in the foking tin, where's the fun in biscuits?

It was said that the Miami Grand Prix had a kind of Superbowl feel to it and the reaction has been really positive. Spain is a very different animal. It's more traditional and you get a lot of petrol heads here who just want to watch racing and are not interested in all the showbusiness thing. I find that quite refreshing after Miami and I love being able to dip into all these different worlds. It makes it fun for all of us.

Anyway, FP1 and FP2. How did it go? Well, it's been a good day today. FP1 was a bit ordinary but by FP2 the boys had found their feet. Mick finished tenth and Kevin twelfth. Not bad. It obviously means nothing if we don't carry it over to tomorrow but I honestly think we will. Mick and Kevin are both really happy at the moment. Perhaps the worm is turning?

Saturday, 21 May 2022 – Circuit de Barcelona-Catalunya, Barcelona, Spain

5 p.m.

The ups and downs of Formula 1 are what keep us going and what make us want to quit sometimes. That's the truth. Elation is sometimes only a hundredth of a second away and the tiniest misjudgement can cause devastation. It's fair to say that so far this season we've had more bad days than good, but today has been one of those days that we live for.

Anyway, how did we get on?

One driver in Q3? Fok off. Not interested. Two drivers? That will do nicely, sir! For the first time since 2019, both our drivers made it to Q3. Kevin in eighth and Mick in tenth. Foking amazing!

Q1 started with a race against time as Mick's car suffered a brake-by-wire issue. It really was touch and go but after some amazing work by the mechanics he got out there and advanced to Q2 with a lap of 1:20.683. Kevin Magnussen had already qualified with a 1:20.227 so, apart from Mick's issue at the start, we couldn't have asked for more.

In Q2 things went smoother and even better overall. Kevin posted a time of 1:19.810, putting him P5, and Mick finished P10 with a 1:20.436. This was the first time Mick had ever made it through to Q3! What a moment. He's got a big smile anyway, but after that you couldn't see the rest of his face. What made Q2 memorable for Kevin was that he had to cope with a malfunctioning DRS system. How the fok he managed to finish P5 is a mystery. That was an intense session!

Q3 began on soft tyres before a final switch to fresh sets for one final time attack. Both drivers managed to improve slightly: Kevin to 1:19.682, which put him P8, and Mick 1:20.368 to seal P10.

As you can imagine, we're all buzzing. The whole team. That's two positive days in a row now. Two! I feel like taking everyone to a bar for a party. I'd better not, though. Let's keep it up!

Sunday, 22 May 2022 – Circuit de Barcelona-Catalunya, Barcelona, Spain

10 a.m.

All the talk at the moment is about upgrades. That's all any of the journalists want to speak about. Some people appear to be fixated by them. 'We need to go faster.' 'OK, let's get an upgrade. It's the only way!' I'm definitely not against upgrades but I don't like using them just for the sake of it. The timing has to be right.

I could be wrong but it feels like they're used for propaganda purposes sometimes. You know, teams making other teams sit up and wonder what they are doing. Then again, it might just be the press and the media feeding the Formula 1 soap opera. Who knows? That seems more likely in my opinion. The upgrades are almost expected, though, which I find quite annoying. 'What, you haven't got an upgrade? But we're in Spain. Everybody has to have an upgrade for Spain.'

The reason I am not interested in any upgrades at the moment is because we haven't yet got the best out of our car and you cannot really upgrade something you don't fully understand and appreciate. That's my opinion and Gene agrees. A case in point is actually this weekend. Every other team on the grid has brought an upgrade, yet our car is one of the fastest. The reason for that is because we have learned more about the car in its current state, including how to lessen the porpoising issue. That is why we've been quicker and that has been our upgrade.

Yet again, the rumour mill has started churning about Haas running out of money, etc. 'No,' I said. 'We just have a different approach to the others. That's all.' Monkey see, monkey do doesn't work with me. It never has. That told them.

I can't wait for the race. I know we had a good qualifying but not having quite as many people and celebrities walking around here is refreshing. There have still been the best part of 300,000 people here over the weekend but, as I said, they're a different crowd to Miami. It's Monaco next week, so by that time I'll be ready for my Hollywood fix again. Lights, camera, Guenther! Don't forget the makeup, though. You'd need a foking bucketload.

6 p.m.

I've said it already, I think, but what a difference twenty-four hours makes. It certainly wasn't the race we were hoping for, that's for sure. Boy oh boy. By the time we got to turn three, Kevin was seventh, Mick was eighth and our pit wall was about ten feet off the foking ground. We were floating to heaven! Then, by turn four, we're one car down and by lap thirty the other is out of contention. With Mick we took a chance with a two-stop strategy and it didn't quite work out, so it wasn't anybody's fault. With Kevin it's a little bit more ambiguous, but that's for another day.

What I wrote yesterday about the highs and the lows of F1 feels pretty relevant at the moment. We've had both extremes in quick succession. The difference now, though, as in here and now, is that nobody is too despondent. We had a clean Friday and an excellent Saturday and although we made mistakes today they were honest mistakes that were down to luck more than anything else. To me it feels like the car and the team are almost at the same point now. We're finally getting to grips with the car's potential so from here on in it should be about refining what we have and then upgrading when the time is right. The team, too, are operating almost to the best of their ability so, with another weekend like this, they should be hitting what we call the sweet spot.

I wish Monaco wasn't the next race, to be honest. As much as I like the vibe sometimes, without some rain it's basically just a parade. As competitive as I think we are now, I don't think we're going to thrive there. I could be wrong, of course, so let's keep everything crossed.

OK, I've got to catch a plane.

Adiós!

Wednesday, 25 May 2022 – Circuit de Monaco, Monaco

2 p.m.

I've literally just arrived from Italy. It's definitely one of my easier journeys. And there are already a lot of people here, which isn't normal. Things usually start getting busier on Thursday. The last week has mainly been about sponsorship for me, which I'll come on to in a second. Before that, I will tell you first what I like about the Monaco Grand Prix.

OK, I'm glad we've got that out of the way. Let's get on to the sponsorship.

I'm only kidding. Well, kind of. Monaco has some things going for it in terms of what the fans can experience in person, but from a team's point of view, not to mention the people watching at home on TV, it doesn't deliver much. No run-off areas, so the drivers cannot take risks. No overtaking. A short, boring track. Until a few years ago there was a rumour that Monaco didn't pay a dime to host the race. That turned out not to be true but the amount they pay is still significantly less than anyone else. Saudi Arabia pay something like $60 million to host their race, whereas Monaco pay about $15 million. Then you have the coverage, which is one of the main things that fans moan to me about. Don't ask me why but, for some reason, unlike every other race, Monaco are in charge of their own TV coverage, which means it's up to them what you see on the TV. I appreciate that it must be quite difficult to direct a race like that, which is even more reason to let a foking professional do it! Last year there was also a big fight about the advertising. For some reason Monaco are allowed to advertise their own sponsors and last year they did a deal with a direct competitor of Rolex, who are one of F1's title sponsors. That, as you can imagine, went down like a big lead balloon.

Their deal with F1 to host races comes to an end this year so it will be interesting to see whether or not they remain on the calendar. It would be a shame to see it go but you have to move with the times and many would argue that Monaco has failed to do that.

Right, I'm even boring myself now. I'll leave the sponsorship news until next time. I'm staying with a friend while I'm in Monaco and I'll be at his place soon.

Thursday, 26 May 2022 – Circuit de Monaco, Monaco

8 a.m.

I had a great night last night. I always stay with the friend I mentioned when I'm in Monaco and he treats me like a king. King Guenther of Monaco. Hotels are OK but being here is a nice change for me, you know. His place faces the sea and after a good meal last night we just sat on his balcony, lit a cigar each and talked shit for a couple of hours. It's not often I get to relax like this during the season and it's done me a lot of good. I'm ready to prepare for the parade now.

The pressure to deliver a new title sponsor is starting to grow a bit but it's all in hand. We're talking to a good number of companies and there's a lot of interest. Because of where Formula 1 is at the moment on the global stage, the value of the sponsorship is high and that has to be taken into consideration. In other words, put a few more zeros on the end! I'm joking. What I'm aiming to do is put something in front of Gene and the board that will help to secure the medium- to long-term future of the team without any drama and without compromising who we are. The drama part, as well as being distracting, is not good for our brand, which is also pretty strong at the moment. A little bit of controversy can be OK sometimes, but I think we've had enough for the time being. We want fewer crashes, less controversy and more points.

The companies we are talking to at the moment come from all over the globe and I've already met a lot of really interesting people. There's a lot to consider, though. Not least, will they be a good fit for Haas? It's not just a case of accepting the highest bidder. I don't want to sound like a marketing guy, but whoever comes on board has to understand our philosophy and be happy with who *we* are. And vice versa, of course. Fortunately, a lot of the interest we have had so far has been from companies who like who we are and like the fact that we aren't as corporate as the other teams. We're not everybody's cup of tea, but that's fine. We're also very good with our sponsors in terms of what they get in return. We're only a small team so they often get to know most of the people. Again, not everybody likes that.

Anyway, I'm off to the track to see how the guys are getting on.

2.30 p.m.

I've just done an interview with Sky Sports UK and the presenter asked me if Haas are going to be able to complete the season. It's something to do with Christian at Red Bull saying that in his opinion certain teams won't be able to complete the season because of rising expenses. Where does Christian get his news from these days? Mick's uncle? There are plenty of teams that hope Red Bull won't be able to see out the season. Ferrari especially.

What Christian is referring to, I think, although in a slightly exaggerated way, *is* actually becoming an issue, as the costs involved in running an F1 team are spiralling. Not out of control, exactly, but the big hard costs such as transport have gone up a lot this year and we have to keep an eye on it. The budget cap was designed to create a level playing field and that in turn has created a bit of a united front. Not about everything, of course. There will still be lots of battles, thank God. That will never change. I do wonder, though, if Christian's 'headline' has anything to do with the fact that we have recently had to submit what we have spent over the past year. Maybe his ass is squeaking a bit?

The final question of the interview I was not expecting. It was something about whether or not Lewis Hamilton should have to remove his piercings before a race. This piece of important news had passed me by, I am happy to say. Jeezoz! I do actually have an opinion, though, which is that if it's in the rules that you cannot wear piercings during a race, which it is apparently, then he should have to take them out. Whether Lewis thinks that's fair or not is immaterial and if he wants to change the rules he should gauge the level of support he'd receive and, if he wants to, start lobbying. I have no piercings and neither do my drivers. Or at least none that they are willing to talk about. You never know, though!

Once again, the number of people here is off the scale. This is quite strange, though, as this year, instead of having practice sessions on the Thursday and then having Friday off, the race is following the standard three-day format. With nothing really happening then, I thought Thursday might be quieter, but not a bit of it. I've never seen it like this before. I assume that it's down to the buzz surrounding the sport at the moment and it's good to see.

Because everywhere is in walking distance at Monaco, we interact a lot more with the fans and I've had some interesting encounters so far. One guy who wanted a selfie was wearing a shirt with nothing but my face on it. Seriously, half his body was completely covered in hundreds of little Guenther heads. It was foking horrifying! Nobody wants to see that, surely? He should have been arrested really, and put in jail.

Friday, 27 May 2022 – Circuit de Monaco, Monaco

8 a.m.

Another easy, relaxing night in Monaco. More cigars on the balcony. More talking bullshit with my friend and host. I obviously haven't told Gertie about the cigars. She'd kill me. Don't worry, there's no danger of her reading my book so my secret is safe. She wouldn't dare! This is the life, though.

A few years ago I got into some troble with the police here in Monaco. My friend's place is up in the hills so I always hire a scooter, and one morning, after I set off down the hill, some kid started racing me. Had I been a normal adult I would have ignored them and carried on at a sensible pace. Unfortunately, I am not very normal so when he passed me I thought, *You want some? OK, I give it to you!* To be fair, he was pretty quick but he was no match for me and when I passed him I gave him a big smile. A few seconds later a policeman appeared out of nowhere and gestured for me to stop. He was really pissed off! Just then my adversary passed us and started waving and beeping his horn. 'Fok you!' I shouted. Surprisingly, that didn't improve my situation with the policeman and within five minutes I was standing in the police station being told off by a sergeant. Jeezoz, he properly tore into me, you know. As this was happening, sensible Guenther was doing battle with idiot Guenther and I had to force myself not to start arguing with this guy. After reading me the foking riot act he said he was going to speak to his superior and then decide whether to charge me or not. 'Charge me?' I said. 'With what? A little bit of speeding? Jeezoz Christ!' That didn't go down very well and, after giving me another bollocking for being rude (me?), he went off to find an electric chair. While he was gone I called Stuart. 'Can you come to the police station, please? I think I might be in a little bit of troble.' He wasn't very surprised. 'OK,' he said. 'I'll come and get you. And please don't say anything else until I arrive.' I did as I was told. That's a pretty normal day in the life of Guenther Steiner.

6 p.m.

It's been a solid day for the most part. Mick had a gearbox issue during FP1, which forced him to stop in the entrance to the pit lane, causing a red flag. You cannot legislate for things like that, unfortunately. The team worked hard between sessions to get him ready and out again in FP2 and he had a good session with no more dramas. The positives are we got a decent number of laps under our belt and in FP2 Kevin finished P11. The pace is there. Hopefully we can unlock more from both cars tomorrow. It's critical to get everything set up for qualifying now and so we'll maximize final practice.

OK, I'm going back to my palace for a cigar and to talk more shit with my host. *Bonne nuit.*

Saturday, 28 May 2022 – Circuit de Monaco, Monaco

5 p.m.

Both drivers progressed from Q1 to Q2, which was good. A few things didn't go according to plan, though, and Kevin qualified P13 and Mick P15. We were hoping to get both into Q3, but there we are. As I keep on saying, we know we've got speed in this car. The drivers are convinced of it. Everyone is. We'll still come out fighting. What we need is some rain, some luck and a red flag at some stage – not involving one of our cars!

9 p.m.

The Netflix guys have been in touch. They want to come out to North Carolina to do some filming at some point. Jeezoz Christ. My wife will be thrilled to bits! One of them mentioned me doing some jet skiing? I said, 'You've got to be foking joking!' They also want to take a look at the composites company I co-own there. Why is that? I hope they're not investigating me. They're good guys and we always have a laugh together. Usually at my expense, though.

Sunday, 29 May 2022 – Circuit de Monaco, Monaco

1 p.m.

The UFC fighter Conor McGregor came to the garage this morning. Fok me, what a crazy guy he is! In a good way. I don't want any repercussions. People think I am an intense guy but Conor is in a different league to me. I suppose you have to be, in his line of work. Do you know what, though? He pledged his allegiance to Haas F1! That's a pretty cool fan to have, you know. The guys in the garage went foking crazy when he came in. In an interview afterwards, he said, 'I'm rooting for team Haas. Our cars are exceptional!' True or not, nobody is going to argue with him. Maybe I should hire him to do our PR!

6 p.m.

Too pissed off to type at the moment. Tomorrow, maybe.

Monday, 30 May 2022 – Monaco

Some rain, some luck, and a red flag that didn't involve one of our cars is what I wished for on race day. Well, I got one of them. Rain! I also got the red flag but unfortunately not the second part. And no luck. None at all.

The reason I didn't write an entry yesterday was because I didn't want to relive what happened so soon. I just wasn't in the mood. I also had a very awkward telephone call to make to my boss.

After a one-hour delay, the formation lap got underway behind the safety car. Kevin then began to climb a few positions and at one point I thought he might be in contention for the top ten. Then, on lap twenty-one, he was forced to retire because of a power unit issue. It was nobody's fault but it was still really frustrating. Reliability has been OK lately.

And now we get on to Mick. Jeezoz, where do I start?

After four laps, he pitted for intermediate tyres, lost track position and then got involved in an incident that left him with a damaged front wing. After pitting for a new wing he went back out and a few laps later he lost control through Piscine corner and crashed. Split his car completely in two. Once again our hearts jumped into our mouths but fortunately he escaped unhurt. That's obviously the most important thing but the fact remains that we have a repair bill totalling almost a million dollars.

The first time a driver writes off a car in a season due to human error, you have to forget about it. It's just one of those things and, at the end of the day, shit happens sometimes. The second time it happens you think, *Hang on, something's not right here.* The cost and the effect it has on our chances of scoring points is one thing, but what about the danger to the driver and to other people? Nobody ever mentions that.

I know I keep going on and on about it but we have a good car this season and I have run out of excuses for things like this. 'He crashed again, Guenther?' the board will say. 'Why? What's the point of having a good car when you don't score any points and keep wrecking them?' What can I say to that? Nothing. Having a good car has quickly turned into a double-edged sword for me and one of the main reasons for this is that we keep on foking wrecking

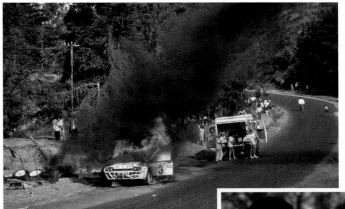

Left: A highlight from my career in rally: standing in my underpants (that's me in white, with the great legs, behind the truck), having helped to destroy a Lancia Delta.

Right: With Colin McRae, the most naturally talented driver I've ever worked with.

Left: Talking to Eddie Irvine. He's crazy, but in a good way.

Below: I couldn't have wished for a better mentor than Niki Lauda.

Above: With Gene Haas, announcing the formation of the Haas F1 Team in 2014.

Right: Contrary to popular belief, Mick Schumacher and I got on OK. He's a good kid.

Above: The start of the 2022 season. Little did we know . . .

Above: 'How about getting Magnussen back?' Gene asked me. 'Do you think he'd do it?' I'm glad he said yes.

Left: Celebrating Kevin's Viking comeback.

Below: First points in Bahrain. Boy, that felt good!

Left: Mick's crash in Saudi Arabia. It's fair to say it was a pretty eventful weekend.

Right: Gene and me looking serious in Miami. I can do serious, occasionally.

Below: Mick and Kevin at the Austin Fanzone.

Above: 'Toto, I have a feeling we're not in Kansas any more.' In Baku, with my favourite Romanian.

Left: Sending a message in Abu Dhabi.

Above: Mick's second crash of the season in Monaco.

Right: 'Our cars are exceptional!' With Haas fan, Conor McGregor.

Left: Top Gunth!

Below: Mick wearing me. I wonder if he still has it . . .

Below, left: Somebody holding up a mask with my face on it, which proves that there are some pretty sick people out there.

Below, right: Sick people at the Canadian Grand Prix.

Right: With Fred Vasseur from Alfa and Mario Isola from Pirelli. Two great guys.

Right: I could make a joke about holding his balls here, but I'm not going to. Kevin preparing in Saudi Arabia.

Below: Mick practising strategy in Canada.

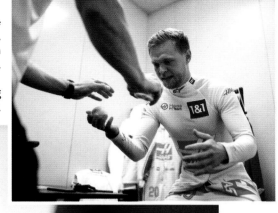

Left: You get asked to do some pretty strange things when you're a driver . . .

Below: Celebrating Mick's first points at Silverstone.

Left and below: Kevin celebrating our first-ever pole position. Hopefully not our last.

Below: Saying thank you to Mick on the last day of the season. I wish him well.

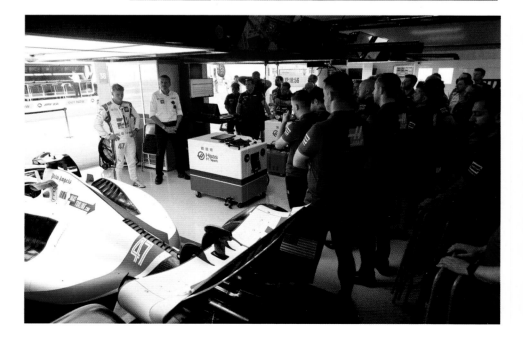

them. Or should I say, one driver does. The driver who hasn't scored any points yet and who is one of only two left on the grid right now.

I'm not saying that nobody else at the team makes mistakes. We all do. This isn't good enough, though. I made a very short statement after the race saying that we need to see how we can move forward from here and at the moment I'm not sure which way it will go. Mick seems unable to appreciate the gravity of the situation, at least publicly, which is also worrying. He talks like it's just one of those things and there is no fault anywhere. If you fok up, admit that you've foked up, apologize, and then try and improve. That's what he needs to do. Although I'd prefer it if he just stopped crashing.

Mick's excuse for the crash was that in order for him to go faster he has to take bigger risks and this was one of the occasions when taking a bigger risk didn't work out. That might be a reason for the crash but it's not an excuse. What does he want? Permission from me to carry on taking risks during races that he can't cope with? He may as well ask me for permission to drive! That's his judgement, not mine. I'm not the one driving the car. How far a driver can push a car depends on their talent and their ability and it's Mick's responsibility to know where to draw the line. He's an F1 driver, for God's sake. He's at the pinnacle of motorsport.

I am employed to run a team because in theory I have a talent for it, and when I have to make a decision I use my judgement. I don't go back to Gene every time and ask for his permission or validation. I also don't use it as an excuse when things go wrong. What I'm trying to say is that if I make a judgement to take bigger risks at Haas and it goes to shit, I do not try and use it as an excuse afterwards.

Gene pays me to run the team, which means I'm accountable for any risks I take. It's part of what I do. We pay Mick to drive a car and he's accountable for that.

I have never seen Formula 1 as intense as it is at the moment, which makes situations like this even worse. Everything is growing. The fan base, the calendar, the coverage. There are still only ten teams, though, and so the pressure to deliver is greater than ever. It's so demanding. What this does, in addition to highlighting our mistakes, is it separates the men from the boys. Or, in driver terms, the ones who can stay on the track and score points for the team and the ones who can't.

I have some serious thinking to do.

Anyway, I want to leave you on a positive. Tomorrow I get to see Gertie and Greta, which is just what I need.

Merci, Monaco. It's been . . . interesting!

Thursday, 2 June 2022 – Steiner Ranch, North Carolina, USA

10 a.m.

I promised to tell you about some stories from my life in rallying a while ago so, as there is no racing this week (or even any foking dramas for a change, which is a first), I thought I'd put them here. Before I do that, I should really give you the story of how I got to work in rally in the first place. My life pre-Netflix, in other words. I did actually foking exist then, contrary to what a lot of people think!

Let's start with school. I was known for one thing at school and that was, of course, talking. Nobody could beat me at that. I was the GOAT. I was also quite clever, though. Surprising, yes? It's true. I was in the top three or four for almost every subject. It didn't interest me, though, being academic. All I cared about was motorsport, despite the fact that nobody else in my family really gave a shit about it, let alone participated in it. I used to watch Formula 1 with my dad on TV a lot but my introduction to live motorsport was a hill climb event that took place about twenty minutes from home. That was what really locked me in, I think.

After leaving school I started an apprenticeship as a mechanic for just road cars and then I had to do my national service. When I finished that I went back to being a mechanic again and then saw an advert somewhere for a mechanic with the Mazda Rally Team, which was based in Belgium. That was in 1986 so I would have been twenty-one years old. I have no idea why they liked me but they offered me the job and so I moved straight over there.

It was a big year for the Mazda Rally Team because, as well as hiring me, they made the jump from Group B of the World Rally Championship to Group A. The following year, in 1987, we led our driver, Timo Salonen, to his first win in the Swedish Rally and the team ended up finishing sixth in the Manufacturers' World Championship. In 1988 we improved again and finished fourth, with Salonen finishing fifth.

My boss at Mazda was the great Achim Warmbold, who is a legend in the rally world. Something strange used to happen to Achim whenever he got emotional or angry. He used to start sweating on his upper lip, which, after a while, made it look like he was foaming at the mouth. It was a very strange thing to see and when it happened I would always make an excuse and run away. I had a good relationship with Achim. Even then I used to speak my mind and at first I think he didn't know how to take me. It's a wonder he didn't sack me really, but in the end he just got used to me, I think.

Although my official position at Mazda was mechanic, I was a bit foking useless and so before I got found out I took a job as the assistant team manager at Top Run in Italy. My main responsibility there was organizing and running private entrants in various rally championships. And we had some success, too. The Belgian driver Grégoire De Mévius finished runner-up in Group N of the World Rally Championships in 1989 and 1990 under my careful stewardship.

Word must have got out about this amazing new talent in the team management industry because in 1991 I moved to Milan to join Jolly Club as the head of reconnaissance for the Group A team. These days a large part of that job would probably be done remotely, but back then you were given a big bag of cash and told to get your passport and just get on with it. And the same with testing, which I was also involved in. That's literally what happened. No cell phones. Just cash, a contact book and a few maps. What an experience.

My King Midas touch continued at Jolly Club when we won the Manufacturers' World Championship in 1991 and 1992, with the Lancia Delta Integrale 16Vs. The following year, in 1993, Carlos Sainz joined Jolly Club and during that year I spent no fewer than two hundred days with him testing and developing the car. Two hundred foking days with Carlos Sainz! Can you imagine that?

That actually turned out to be one of the most important times in my career. As well as being an incredible driver, Carlos Sainz is one of the most professionally driven human beings I have ever known and he never misses a beat. That rubbed off on me pretty quickly. I had no choice really. It was either follow his lead and try and be the best or get the fok out! I also learned a lot about discipline and about how to conduct yourself from Carlos. Most importantly, though, I learned that by working hard and having the confidence to push yourself forwards you can achieve something in life. Carlos is over sixty now but when he does the Dakar Rally he is still the first one up and the last one to go to bed. He's relentless.

By 1994 I had been promoted to technical manager and was responsible for the two Ford Escort RS Cosworth cars that competed in Group A. Yet again, success followed me around like flies on shit, as not only did we win back-to-back Italian Rally Championships in 1994 and 1995, but in 1996 we won it again!

In 1997 Gertie and I fancied a change of scene and so we moved to Banbury in the UK where I became the team manager for the Prodrive Allstar Rally Team. The team was brand new yet we still managed to win the European Championship in our first season.

You thought I was some kind of foking no-mark chancer who didn't know what he was doing, didn't you? Admit it. It's not true, though. In some cases, I even know what I am talking about.

After one year with Prodrive I was asked by my friend, who later became my mentor, Malcolm Wilson, to join his team M-Sport as a project manager for the team's WRC Ford Focus fleet. Soon after joining, Ford chose M-Sport to design and build its new Ford Focus World Rally Car. I cannot take credit for that one, though. That was Malcolm. I then began taking more of a leadership role to establish technical facilities and assumed overall management of the car's development.

The new car appeared the following year, 1999, with Colin

McRae driving, and immediately grabbed headlines for recording the fastest stage times during the Monte Carlo Rally, which was the team's first ever event. M-Sport scored its first World Rally Championship victory in its third event of the season at the Safari Rally in Kenya and a month later Colin scored his second win in a row at the Rally de Portugal. He ended up finishing sixth in the Drivers' Championship and the team finished fourth in the Manufacturers'. Not a bad start.

I should take the opportunity to say a few words about Colin here, as he is by far the most talented driver I have worked with. He was the opposite to Carlos in that he didn't need to work at it very much. His talent was purely natural and that could create a pretty unpredictable environment. He did what he wanted, basically. Not in a bad way, but you had to be ready for anything. He was definitely at his best as a driver when there were not a lot of electronics or driving aids in the car. It was all about control with Colin, and the more control he had over a car the better he became. Testing with Colin is the best example I have for this. He was either in the mood for it or he wasn't, and if things weren't going his way you would know about it. Then again, if things *were* going his way you were in for a foking treat because he could make the car do anything.

I've often been asked if I think Colin could have become a successful Formula 1 driver and it's an easy one to answer. If Colin McRae had *wanted* to become a successful Formula 1 driver there's no question in my mind that he could have. Would he have wanted to put the work in, though? That's the question. All Colin wanted to do was get in a car and drive, so the distractions that inevitably go with driving an F1 car would probably have put him off.

Following the 1999 World Rally Championship I was promoted to director of engineering at M-Sport. Then, in 2000, I worked with Colin McRae *and* Carlos Sainz Sr to finish second in the Manufacturers' World Championship, while they finished fourth and third respectively. In 2001 we finished second *again* in the Manufacturers' World Championship while Colin finished second and Carlos sixth overall. Soon after that I was approached by Niki to join Jaguar.

I don't want to go on too much about my rally days as this is not an autobiography or a book about rally. I still think that a couple of stories might be fun. Do you know what rally gave me, though, apart from thousands of experiences and memories? An understanding of motorsport and how it works from the bottom up. I don't see it from the top down. I never have and never will. I can tell people what to do from the top down. Sure I can. But that is not where I see it. The bottom is getting the car out on to the track and that is what I learned to do first. I obviously don't know everything, but I can speak with a mechanic the same as I can speak with a CFO or a production manager. It's very difficult to fool me and tell me bollocks, you know. And that's mainly down to rally.

One of my favourite stories from my rally days happened in Kenya one year during the Safari Rally. I was working for Jolly Club and we were running a Lancia Delta. Me and another guy were in charge of the refuelling and were both wearing fireproof overalls. There was this foking great big aluminium tank and while one of us controlled the bleed valve the other poured the fuel in using a jug. It wasn't very hi-tech and it was also very dangerous.

I don't know how it happened exactly, but as we were refuelling the car some fuel managed to escape and, without any warning whatsoever, the entire car suddenly became engulfed in flames. You know the sound that makes when you see it at the movies? It was exactly the foking same. Whoosh! It could have been static that ignited the fuel or it could have been the brakes. Either way, we had a fire on our hands! The first thing that me and my colleague did was drop the fuel can, which didn't really help matters much. How many gallons it had in it I'm not sure but it was enough for us to toast a few million foking marshmallows.

One of the first things I noticed after we dropped the can was that I was now on fire, too. My overalls were fireproof, in that they prevented fire from penetrating the material, but the material itself was still on fire.

'Holy shit!' I said. Or maybe screamed. I then proceeded to get the overalls off as quickly as I could. The only thing I had on underneath them were my underpants and when the overalls were off I started jumping up and down on them to put the fire out. Luckily, a very famous rally photographer called Colin McMaster was on hand to take a photograph of me doing this and if the publishers have done what they were threatening to do it might be in this book somewhere.

Once I'd put my overalls out, which was obviously the most important thing to consider when the car was like a fireball, I suddenly remembered that there were two people still inside the car. Fortunately, the driver had already managed to free himself but the navigator was having troble undoing his belt. Thinking nothing for my own safety, or for the fact that I was only wearing a pair of pants, I went to help the navigator get out of the car. What a hero, eh? Guenther to the rescue. Once that was done I heard a radio going. It was somebody in the helicopter above.

'What is happening down there?' they asked.

'What is happening? We're all having a nice foking safari picnic. What do you think is happening?'

I remember looking at the scene just then and it was madness. Somebody was in the process of trying to move the service truck, which had all the fuel in it, and there were people running round in various states of panic and at various stages of undress. It was just crazy.

One of the people helping us had been quite badly burned in the accident so I told the person flying the helicopter to land as quickly as possible and then take this guy to hospital. He was making a lot of noise but nobody knew what to foking do.

The helicopter and its crew eventually landed and when they'd gone again and we knew that everyone else was OK, an immediate inquest started into what had happened. 'Look, it's nobody's fault,' I said. 'It's just one of those things. These tanks weigh a ton and it's easy to spill fuel sometimes.'

Finding somebody to blame wasn't on my agenda and everybody else felt the same. Except for one person. 'I think it was Guenther's fault,' said the guy who I'd been refuelling with. 'What the fok? Come on, man,' I said. 'It was just an accident. We are not apportioning blame here.' I couldn't believe it. This guy wasn't having any of it, though. As far as he was concerned the accident was my fault and that was the end of it. It was nothing to do with him. 'There were two of us doing the refuelling,' I said, trying to stick up for myself. 'How can you be sure that it was me?' He couldn't give me an answer but that didn't change how he felt. As far as he was concerned, Guenther was the fire starter. The twisted foking fire starter!

OK, I've got one more story I will tell you, but not today. It'll take too long.

Friday, 3 June 2022 – Steiner Ranch, North Carolina, USA

12 p.m.

I went to bed last night thinking about the story I am going to tell you now and I ended up having a dream about it. It's completely foking crazy, but I promise you it's true.

We're going back thirty years – to 1992 for the Dakar Rally, which, that year, was Paris–Cape Town instead of Paris–Dakar. Don't ask me why, I have not got a clue. I'd got a job as a co-driver for a service truck for one of the teams and I was seriously looking forward to it. I'd heard all kinds of stories from my colleagues in rally about the Dakar but had never been involved in any way. 'Put it this way. It'll be a foking adventure,' one of them said.

One of the first things I learned about the Dakar Rally was that, as well as being one of the most famous motor races on the planet, it was also one of the biggest insurance scams, too. I cannot say who or when because I don't know, but apparently teams used to take old race cars and then burn them out in the desert and claim on the insurance. It was famous for it.

The truck that I was co-driving had once belonged to the Monaco royal family and had been customized accordingly. It had awnings that were brightly coloured and all kinds of other shit on it. The team must have got it cheap, I suppose. Within three hours of setting off from Italy to Paris we caused havoc when the turbo charger on the truck blew. All the oil started running through the exhaust and out on to the road and within a minute there were cars sliding foking everywhere! Somehow we managed to stop the flow of oil and get the turbo charger changed without anybody being killed. Then, after the rally started in Paris we made our way to the next finish point, which was Sète on the south coast. From there we drove to Marseille and took a cargo boat over to Libya, which is when I started to realize that the team I was working for might not be quite as well funded and professional as I'd been led to believe. The boat they'd booked us on was barely even seaworthy and appeared to have no mattresses or cushions anywhere. Everywhere was just steel! When we disembarked the following day, not only could we not walk properly because none of us had slept, but we were all ready to kill somebody. Anybody!

'Well, at least it can't get any foking worse,' I remember saying to my co-driver.

'Really?' he said. 'Have you ever been to Libya at this time of year?'

I'd assumed it would be boiling hot and dry but I was wrong. It was raining and it was cold.

'Where the hell do we sleep?' I asked.

'There's a makeshift village for the rally teams a few miles away,' he said. 'Some stay in motorhomes but we'll be staying in tents.'

'Jeezoz Christ. Really? I've just spent a night on a bed of steel and now I have to sleep on the foking ground?'

'Apparently there are tents in the truck somewhere.'

He then gestured to the third member of our three-man Italian team – a guy in his mid-fifties who was as dumb as fok – to go and find the tents. He came back about ten minutes later with a big bag and we assumed that in the bag were three tents and three sleeping bags.

It didn't take us long to find the team village and, sure enough, it was a case of the haves and the have-nots. Or, the motorhomes and the tents. We found a space that we thought would be big enough for three tents and then emptied the bag.

'What the fok?' I said. 'That isn't even one tent, let alone three. And where are the sleeping bags?'

What the team had given us were three blankets and an awning. Not a tent. An awning. The kind of thing that British people put up in their gardens in summer because they know it will foking rain at some time.

'What are we going to do?' I said. 'I'm not sleeping under that foking thing. No chance. I think we should find a hotel.'

With no arguments coming from the other two, we set off on foot to try and find somewhere to sleep that might actually keep the rain off and be warm. We'd literally been walking for about two minutes when we came across a guy standing at the corner of a street holding an AK47. It's not so strange in Libya. Or at least it wasn't then.

'Do you speak English?' I asked.

He nodded.

'Any idea where we can find a cheap hotel? One with a foking roof, preferably.'

He nodded again.

'OK, then. Could you take us to it, please?'

'Only for beer,' he said quietly.

Alcohol was illegal in Libya then but we had a ton of it in the truck.

'I want two beers. Then hotel.'

'OK, then,' I said. 'Follow me.'

We took this guy back to the truck and gave him two cans of beer, which he drank in about thirty seconds. Halfway through the first can he threw his AK47 on the ground. 'What the foking hell are you doing?' I shouted. 'You could have foking shot one of us!' He was too busy drinking beer to give a shit about what I was saying and said nothing.

I was half expecting him to ask for more, but after the second can he took us straight to the hotel. Before he left I asked him if he could come back in the morning and take us to the start of the rally. 'Sure,' he said. 'I'll see you tomorrow. More beer, though.' I didn't think for a moment that he'd turn up but at seven o'clock the following morning he arrived at the hotel carrying his AK47.

'Can I have beer?' said the man.

'What, for breakfast? Fok me! Sure you can have a beer. You can have one while you direct us to the start in Misrata.'

By the time we arrived in Misrata for the start of the next stage he'd had three beers and was as happy as a pig in shit. We paid him off with four more and some money to get home and then sent him on his way. During the next stage, which was to the city of Sirte, we developed an issue with the fuel pump and had to do a botch job that involved us pumping the fuel from the top of the truck into the fuel tank. After that the route master broke, which meant the only thing we had to guide us to where we were going was a foking compass! We were four days in but it had already become one of the most foked-up things I'd done in my life.

While driving through the desert towards Niger, we discovered that the old guy, who also drove the truck sometimes, could drive while he was asleep. I'll never forget this for as long as I live but he would sit there upright while holding the wheel, close his eyes and fall asleep while maintaining exactly the same speed and direction. He even used to start snoring! It only worked in the desert, though, as you rarely saw another vehicle.

I forget when this was exactly but while we were making our way to Niger the first co-driver, who was about my age, suggested we take a short cut.

'But we're in the desert,' I said. 'How can you know a short cut in a desert?'

'Trust me,' he said. 'I've been here before many times. I know where I'm going.'

I didn't believe the idiot but what could I say?

About ninety minutes after taking this short cut of his, something weird started happening to the truck.

'What the fok is happening?' I said. 'Jeezoz Christ, we're sinking. We are, we're foking sinking! Come on, what's happening?'

'Erm, I think we might have driven into a salt lake,' said my co-driver.

'A foking salt lake? You have to be joking. What is this a short cut to? An early death?'

We must have been travelling at about 80 kph when we started sinking and we went down about 2 metres. I'd never experienced anything like it before.

Within half an hour about fifty locals had gathered around the truck and we managed to pay a few of them to dig us out. In the end it took them two whole days! We were so behind. While this was happening the sweeper (that's the name of the person who looks after everyone outside the rally) offered to take us to the nearest airport but we refused. Had we done that we'd have had to leave everything there and we couldn't. Instead, we decided to wait to be dug out and then carry on driving. 'We stick to the foking route this time, though,' I said. 'No short cuts!'

To make sure we didn't get lost again we hired one of the locals, who had offered to guide us until we caught up with the rally. He was about the same age as our sleep-driver but was suffering from the same personal hygiene problems we were and obviously hadn't washed for a few days. This meant we now had four stinking bastards — three Italians and a Libyan — in one cab. It was just disgusting. We tried keeping the windows open but it seemed to make no difference.

God knows how but we managed to get to Niger and although we were behind the rally still, we carried on. It's amazing that we hadn't fallen out with each other yet. The closest we came was when somebody farted, so about every twenty seconds! There'd be a shouting match that lasted about a minute and then everybody would stop suddenly. We were all having about three or four hours sleep a night so were probably too tired to fight.

Just as I started to allow myself to become slightly optimistic about our chances of catching the rally and clawing back a bit of normality again, a sandstorm started that stopped us dead in our tracks. We were somewhere in between Dirkou and N'guigmi and you couldn't see more than about half a metre in front of you. It was crazy! We just sat there like a bunch of wankers waiting for the storm to pass. Night came so we tried to get some sleep but it was impossible. Four already stinking men in a cab with no air. Are you kidding me? It was torture. When morning finally came the sand-storm had calmed down just enough for us to open a window a bit to let some air in.

'I can hear camels,' said the smelly Libyan after opening it.

'What?'

'I can hear camels!'

Before I could call him a prick and tell him to shut up he jumped out of the cab and ran off into the distance.

'Where the fok are you going?' I shouted, but he didn't stop.

Three hours passed, by which time I had convinced myself that the Libyan must have been hallucinating and was probably lying dead on a dune somewhere. I was just thinking about how we might get word to his family in Libya when the driver's door opened. It was him.

'Where the foking hell have you been, you prick?' I said.

'I told you I heard camels,' he said. 'And I did. Look.'

He turned around and pointed and there in the distance were four men on camels.

'Tuaregs,' he said. 'They will help us.'

By this time I thought I must have been hallucinating! It was so foking weird.

After a bit of negotiation, one of the Tuaregs offered to get in the cab and guided us to their tents, which meant we now had five stinking bastards in there. It took about an hour to get there and when we arrived they gave us something to drink.

'What is this?' I asked. It looked disgusting. Like seriously off milk.

'Camel yoghurt,' said one of the Tuaregs.

We couldn't really refuse it as they had been so kind to us, so we had to drink it. It was horrible! I thought, *If these guys don't end up killing us, this stuff will.*

We ended up doing a deal with the Tuaregs where they would lead us out of the sandstorm and to an army base where we could refuel the truck. Two of the Tuaregs insisted on coming with us and if it hadn't been for the Libyan deciding to cut his losses and go home it would have been unbearable. The journey still lasted five hours. I have quite a good sense of smell and every thirty seconds or so I would catch a waft of our collective stink and start retching.

By the time we arrived at the army base we'd escaped the sandstorm and so had been able to open the windows for a bit. It was still pretty disgusting, though, and so after refuelling we asked somebody if we could have a shower. He looked at us and shook his head, as if to say, 'What, let a bunch of dirty fokers like you into our showers? No way!' I don't think I'd ever been as disappointed in my entire life. I couldn't escape my own foking smell, let alone the other two guys'!

The following day we finally found somewhere to have a shower and then had two uninterrupted days of driving with no disasters. At last, our luck was changing. On the third day we started running low on fuel again and so started thinking about finding somewhere to top up. We pulled over to talk about this and after literally a couple of minutes an army truck with a fuel tank on the back pulled up behind us. We were in the middle of the foking desert! Me and the younger co-driver got out of the cab and went to talk to the driver.

'Can we buy some fuel from you?' we said.

'No,' said the driver.

Just then I noticed that on the front of the truck was an antelope

that they must have shot. It was quite a common sight, though, so I
didn't think much of it.

'You can buy some fuel if you buy half the antelope,' said the
driver.

'What the foking hell would we want with half an antelope?' I
asked.

'What about the whole antelope?' asked the driver.

'Well, if it'll get us some fuel, why not?'

After filling up we went on our way again with a dead antelope
in the back of the truck. The only thing that was pissing me off now
was the fact that my lips were hurting. In fact, they were killing me.
As well as the weather being hot I was pretty dehydrated and they'd
been chapped for several days. 'This is seriously painful now,' I
remember saying to my co-driver. Apart from finding a drug store
or a supermarket – in the foking desert – there was nothing we
could do. I just had to grin and bear it.

About an hour later we were driving through a small village
when in the distance I saw somebody driving an ATV, which is
what Americans call quad bikes. 'Hang on,' I said. 'I know that
guy.' It was a friend of mine who was part of the service crew for
another team. 'What the fok is he doing here?' I said. 'I'm going to
go and speak to him.' It turned out that he had got lost so in return
for helping him find his way again he gave me a tin of lip balm. I
could have kissed the guy!

Even though we knew where we were going, we hadn't managed to catch the convoy yet. This presented a big problem, as Round
10 and 11 were supposed to go through Chad, where there was a
war on. The convoy with the rally had been granted permission to
travel through the country but there'd been no word on stragglers.

'It'll be OK,' said the guy in the ATV. 'They'll know that we're part of the rally.'

'And what if they don't give a shit?' I said. 'If they kill you, nobody will ever know. And even if they don't kill you, they'll take everything you have.'

Despite what I said, the guy with the ATV decided to go it alone through war-torn Chad. I tried to persuade him not to but he was adamant. 'I want to catch up with the rally,' he said. 'I'll be OK. You'll see.' What an idiot! Me and my two colleagues ended up contacting the owner of our team, who told us to drive to Benin on the west coast, put the truck on a boat back to Italy and then find an airport and fly back home. It took us a few days but we didn't care. Funnily enough, we ended up arriving back in Italy about the same time as the drivers and the rest of the team, but when we told them what happened they didn't seem impressed. 'It's Dakar,' somebody said. 'It's what happens.'

About a month after the rally had finished, I caught up with the ATV guy and when he saw me he went bright red. 'You were right,' he said. 'Don't tell me,' I replied. 'They took everything?' 'Yes,' he said. 'The lot.'

What an idiot.

I've just realized that I have spent ages writing down two rallying stories for this book that actually has no foking rallying in it! Now who's the idiot? Jeezoz.

Monday, 6 June 2022 – Steiner Ranch, North Carolina, USA

11 a.m.

I went out earlier and I was stopped by somebody who, after getting a selfie, asked me what it was like being famous. I think I replied that it was OK, but it actually got me thinking. It's something I've never been asked or have even thought about before. So, what is it like being famous? Well, it can get pretty crazy sometimes. When I arrived at the airport to fly back to Charlotte from the Monaco Grand Prix a couple of weeks ago, a woman asked me for a selfie. 'No problem,' I said. I then realized that she was wearing a T-shirt with my foking face on it. 'Jeezoz Christ,' I said to her, 'You need help!' It's not the first time that's happened. In fact, the last time was actually at Charlotte Airport. My daughter was with me and once again I was approached by somebody for a selfie who was wearing a T-shirt with my face on it. My daughter was totally horrified and I could tell what was going through her head. *Why on earth would somebody want to have a T-shirt with him on it?*

I'm not actually sure how Greta feels about her dad being famous because we never really talk about it. I've wanted to find out sometimes but I always think that as soon as I mention it she'll think I'm being pretentious. The only thing that used to confuse her sometimes was when people started saying hello to me. Before it all happened the only people who said hello were people who knew me, whereas after *Drive to Survive* lots of people started doing it. She just takes it all in her stride, I think. It all happened very gradually, so over the years we've got used it. I'm a superstar now. Get over it!

I'm not even sure how I feel about being famous, to be honest. Before *Drive to Survive* started I would turn up at a Grand Prix and hardly anybody would say anything to me. Apart from, 'Who's that weird-looking shit over there? Surely he cannot be in Formula 1?' Now when I turn up it can take me half an hour to get to the paddock. Not everybody's pleased to see me but if the people who are want to chat and have a selfie, that's fine. The only time it becomes a problem is when I have somebody like Gene with me. I can't keep him waiting while I pose for selfies so if he's there I whisper, 'I'm with the foking boss!' and walk on.

The people who have the most fun with it are the drivers and the mechanics. The mechanics are always taking the piss out of me. Sometimes when the grid walk takes place they'll push me out of the garage on to the grid like some kind of freak show. 'Come and look at the silly Guenther! Why don't you poke him and call him names!' I'm like some kind of foking mascot. I take it all with good humour, though. After all, I'm not the Pope. He wouldn't be able to cope with running an F1 team. Then again, could you imagine me walking on to the balcony in St Peter's Square in Rome and delivering the Urbi et Orbi. 'OK, I'm finished now. Get to foking work!' Pope Guenther the first would be Pope Guenther the last!

I've got a meeting soon with our marketing director about the new title sponsor and then later tonight I fly to the UK and then on to Baku. We're at a stage now with the sponsor when we have to make a decision about who we go with. Having more than one company that wants to work with us is fantastic but we can't keep them hanging around. We have to reach the hallelujah stage, as I call it. We're just balancing everything up at the moment. What they all want, what we can give them, etc. The decision will have to be made soon, though, otherwise we will lose credibility. The most important thing to me about this deal apart from the fit is the longevity. Not just the length of the initial contract but trying to make sure they'll renew in a few years. We've had other title sponsor issues over the years apart from Uralkali and I don't want it hanging over us any more. Enough already!

OK, time for my meeting.

God be with you, my children.

Thursday, 9 June 2022 – Team Hotel, Baku, Azerbaijan

7 p.m.

I promise you I'm not on the Azerbaijan Tourist Board's payroll but if somebody who was new to Formula 1 asked me to recommend an exciting Grand Prix in a nice city that isn't too crowded, I would say Baku. The city is beautiful, the people are great and the track is very fast. The drivers love it and so do the fans.

It started the same year as us, in 2016, but for the first year it was known as the European Grand Prix. If Baku was in the middle of Europe you'd have hundreds of thousands of people here every year but at the moment they get about ninety thousand. It's growing, though.

From my point of view, the fact that it isn't too busy is excellent as it means we have far fewer distractions and can concentrate more on the racing. Not that having more people around is a problem for us. At least normally. The quieter events are great, though, and especially ones that suit our car and deliver a good race.

The only real issue I have with the Azerbaijan Grand Prix this year is that it's back-to-back with Canada, which is 5,000 miles away on a different continent. Every team is in the same boat but it's going to be a killer. If you have any damage on the car, it gets even more difficult so we're hoping to have a clean weekend. For next season it would be great if we could combine the races regionally, or as much as possible. I know that Stefano is working very hard on that one at the moment and hopefully it will improve.

I wish I had a dollar for every person who has said something to me like: 'It must be great working in Formula 1. All that travelling!' If I did, I wouldn't just be able to poach Adrian Newey – I'd be able to buy him from his family and keep him in a foking box! A lot of people assume that Formula 1 is only ever glamorous and that everybody flies first class or on a private jet. Not true. About 60 per cent of our staff fly economy.

The whole glamour thing is just a front really. A myth. Sure, there's a lot of money involved in F1 and some people are paid a fortune. I cannot think of another sport, though, where the people behind the scenes work as hard as they do in F1. The life of a mechanic, for instance, is anything but glamorous. These guys work their asses off and the responsibility that sits on their shoulders is enormous. I think this might have appeared in *Drive to Survive* but a few years ago a mechanic of ours didn't put a wheel back on correctly during a pit stop and it ended up costing us points. The mechanic concerned was distraught but there's no escaping the fact that it made a difference to our season. A lot of the time these guys will be jetlagged and exhausted but you have to rise above that. I wouldn't do it for anything.

So, what do I think our chances are here? The circuit is very fast and very twisty and that could suit us. We need to hit the sweet spot on the set-up and I think we can have a good result like we could or even should have had in Spain and Monte Carlo. I'm really looking forward to it.

Friday, 10 June 2022 – Baku City Circuit, Baku, Azerbaijan

9 a.m.

The mood in the camp is good at the moment. Despite what happened with Mick, the take-home memory from Monaco for the majority of the team, apart from meeting Conor McGregor, has been what could have happened with Kevin. He could easily have finished ninth in that race and everybody knows it. For me it's been slightly different and, although I'm in agreement regarding what Kevin might have achieved, the crash has caused me a lot of headaches. It's not only the budget. It's also the production of the parts. You've only got one or two sets of moulds, and you cannot make more. It's very difficult to keep up at the moment.

The majority of the team arrived here on Monday and because the freight didn't arrive until Tuesday they had a little bit of downtime. That doesn't happen very often during a double header. Also, after Canada next week the next race is Silverstone.

As well as getting some time at home with their families, some of the team will even be able to sleep in their own beds over the race weekend. That's a seriously rare thing for somebody who works in Formula 1.

6 p.m.

FP1 started badly with a water leak on Mick's car, which meant he couldn't go out. That isn't good at a track like this. I don't know what happened exactly but we fixed it for FP2. We didn't perform as we planned in FP2 so we need to go back to the drawing board overnight and see what we can do for tomorrow. The performance just isn't there at the moment.

Anyway, I'd better see if I can get some rest. I've got a team principal press conference tomorrow with Mattia and Toto and for that I will need to be at least half awake!

Saturday, 11 June 2022 – Baku City Circuit, Baku, Azerbaijan

The press conference went OK, I think. I tried taking the piss out of Toto a few times but he didn't bite. That was disappointing as, in addition to dishing out bullshit, giving people bad news and persuading people to do things, it's something that I'm usually pretty good at. I won't give up, though. Next time I'll get him.

One of the journalists asked about porpoising, which is still an issue for most of the teams. Apparently one of the Mercedes drivers (I forget which one) has said that if it continues we need to rethink the concept. I think it's a bit early for that. We're only a few races in and the engineers will probably fix it. I think we need to wait a little bit longer. If we can't, a time will come when we'll need to talk about changing the regulations. In general, though, the cars are not working badly.

5 p.m.

Because of all the hard work the guys put in overnight we managed to get some of the performance in FP3 and we were actually looking forward to qualifying. Unfortunately, everything went against us and we didn't make it out of Q1. On our last attempt to get out we were too late and so not allowed in the queue. We keep saying we're unlucky and at some stage it will change but today we were our own worst enemy. Not nearly good enough.

Sunday, 12 June 2022 – Baku City Circuit, Baku, Azerbaijan

6 p.m.

Jeezoz Christ! You know what's coming, I suppose? That's right, another weekend when we should have left with something but didn't. I'm starting to sound like a broken record now. What can I do, though? Mick finished P14, Kevin didn't finish, and Guenther is now banging his head against a wall.

Anyway, there is something else I need to talk about. I have to get this off my chest. Just as I was leaving the paddock to go to the pit wall before the race I was asked by Sky Sports Germany to do a quick interview. I assumed they wanted to ask me some questions about the race and so I said yes. Instead they started stirring up a load of shit about Mick. They said that according to his uncle I don't speak with Mick and I don't help him. Really? That's news to me. The guy interviewing me, Peter Hardenacke, who I usually get on OK with, was quite confrontational and I must admit that it took me by surprise at first. Then, after a few seconds, I found my anger and I came out fighting. What they're trying to do by making these accusations is to divide the team and create headlines. As well as being annoying I think it's very unprofessional and has nothing to do with Mick. Anyway, Peter got the surprise of his foking life when I turned on him.

I was talking to Johnny Herbert the other day, who I know from my days at Jaguar, and he said to me that in his eleven years in Formula 1 the only driver he ever took against was Mick's uncle. Johnny Herbert is a very tolerant human being so you have to be pretty special to get on the wrong side of him. He's also a very good presenter and can be critical and controversial without being personal. Mick's uncle isn't smart enough to understand the difference, I don't think. Anyway, I gave as good as I got in the interview and next time I'll be ready for them.

Wednesday, 15 June 2022 – Team Hotel, Montreal, Canada

Although I'm exhausted it's nice to be back in Canada. I say nice because the people of Montreal are always really pleased to see us.

It's a great event for their city and the entire F1 community loves Montreal. It's also a good race more often than not and I love the atmosphere.

I'm expecting the porpoising to be bad here. Although we've made progress, we still suffer a little bit. It's different from circuit to circuit. In Barcelona, it seemed like everyone had found a solution to it but in Baku we were back to square one. We'll know more when we go out in FP1. Canada is pretty bumpy, so if it's bumpy it probably won't get any better. But let's see. The team principals are due to have a meeting about it sometime over the weekend and one or two people might get a little bit emotional.

Hang on, I've just written an entire entry without any complaining or any swear words. Wow!

Friday, 17 June 2022 – Circuit Gilles Villeneuve, Montreal, Canada

FP1 wasn't ideal but it was still a good session because we learned a lot. FP2 got us closer to where we want to be. We're still not there, but at least we didn't encounter many problems and there was very little porpoising. Now, let's see what comes tomorrow. The weather could make a big difference.

That's my second entry without swearing or complaining. It won't last for ever. It feels wrong, somehow.

Saturday, 18 June 2022 – Circuit Gilles Villeneuve, Montreal, Canada

1 p.m.

The team principals' meeting took place earlier regarding porpoising. Boy oh boy! It was very interesting. Toto and Mercedes are arguing that it affects driver safety and want the FIA to change the regulations. Christian, on the other hand, has accused Mercedes of having designed a car that makes the porpoising worse and thinks that they should fix it rather than ask the FIA to change the regulations. There have been rumours that teacups were thrown during the meeting but that's bullshit. Toto did get quite animated at one point but that's normal. Netflix were there so those guys will be celebrating tonight, for sure. That white-collar boxing idea of mine might still happen one day.

7 p.m.

If my driver who took me to the track this morning had said to me, 'I'll tell you what, Guenther. Today Haas will block out the third row of the grid in qualifying,' I'd have assumed he was taking the piss. Not because I don't think it's possible, but because of the luck we've been having lately. Or should I say, the lack of luck. Anyway, that's exactly what happened. Kevin qualified P5 and Mick P6, which is our best ever two-car qualifying. I'm trying not to get too carried away now but it's difficult. The mood in the team is pretty good at the moment but there's a long way to go.

Sunday, 19 June 2022 – Circuit Gilles Villeneuve, Montreal, Canada

7 p.m.

You really want me to tell you about the race? Come on, you know what happened. Don't make me do it! OK, then, you win. Both drivers got a good start and for the first two turns they held their positions well. Then, at turn three, Kevin got cocky and decided to make a move on Hamilton. Even if he'd managed to pass him, the Mercedes is a stronger car with better race pace so Hamilton would have got his position back. He didn't, though, and during the manoeuvre Kevin made contact with Hamilton and so that was the end of his race. It was a bad decision, and especially so early on. I think he finished seventeenth in the end.

Mick also had a really shit afternoon but through no fault of his own. He was running well until lap eighteen but then had an engine failure. The only positive is that it's been his strongest weekend to date and had he not had to retire he would probably have been in the points. As it is, we've scored two points in the last seven races. Please, give us some foking luck for a change!

Wednesday, 22 June 2022 – Castello Steiner, Northern Italy

As much as I love spending time in Italy, I don't have an office here. This means I have to work in the living room and I foking cannot do it. I have to do so much talking with work and sometimes it can get very intense. If other people are here it's a distraction. Especially Greta. She's only thirteen so when she's around I have to change from full-fat Guenther to semi-skimmed Guenther. Sometimes even skimmed Guenther, if she has a friend here.

The architect who is supposed to be designing my office was here at 8.30 a.m. this morning, which is the middle of the night to those people. It was supposed to have been sorted months ago but he's always on holiday. I said, 'Dude, come on. This was supposed to have been done by now.' Nobody wants to work any more. Jee-zoz, I sound like a grumpy old man. You can see I'm having a good day already.

I fly to England later today and my first stop will be the factory at Banbury. I've got a lot of meetings planned but it will be good to see the guys there. I said earlier that it isn't always easy for me to visit so I'm looking forward to getting in the way and distracting them from their work.

Next Tuesday we will be holding a family day there for the team. Events like these are very important. Not only because the staff get a chance to get together in a social environment for a change and show their families what they do for a living, but also because we get to say thank you. Until Covid hit we used to hold one every year. Then, in 2021, when things were returning to normal, we decided to hold a Christmas event instead. That one had to be cancelled at the last minute due to a big surge in cases so it's great that we can finally get something going again. The guys have got all sorts of things planned, and Kevin and Mick are coming too. It's going to be a lot of fun.

Silverstone is a special race and most people in F1 think the same. For sure, the crowd like to have fun, but they all know their motorsport and are very enthusiastic. I have no idea how many fans I speak to while I'm there every year – hundreds, probably – but I genuinely enjoy meeting them. I also feel embarrassed sometimes by how much knowledge they have. It's more than me, that's for sure.

But it's not just the fans who get the balance right. So do the organizers. Given the history of Silverstone and the British Grand Prix it would be very easy for them to concentrate too much on the past and ignore the present. They don't, though. They are respectful to the past but they big up the present and the future. As a motorsport fan, I always look forward to what they have planned.

I was going to try and write something about how I think we might get on at the Grand Prix but I can't think that far ahead yet! I'd also be tempting fate, perhaps. Best to keep my mouth shut.

Thursday, 30 June 2022 – Silverstone Circuit, Northamptonshire, UK

5 p.m.

I must have spoken to at least a hundred spectators today. Seriously, these guys are incredible. I enjoy meeting people who want to talk about Formula 1 in its present state, but when they start talking about historical stuff I get really engaged. I wouldn't say I'm an F1 nerd exactly but I am definitely an enthusiast. The difference between Silverstone and other races is that here I sometimes end up asking the spectators questions. It's crazy.

Friday, 1 July 2022 – Silverstone Circuit, Northamptonshire, UK

8 a.m.

What the hell is happening to Bernie? I've just seen an interview with him on TV and after being asked if he was still friends with Putin he said he'd take a bullet for him. What the hell?! He then said that Vladimir Putin was a first-class person. You're a ninety-one-year-old billionaire, Bernie. For God's sake go and buy an island or something.

3 p.m.

FP1 was a washout for everybody really and only ten teams recorded lap times. FP2, though, was pretty good for us. We got all our laps in, did all our work and we know what we've got to do for tomorrow. Dare I say that I'm quietly confident? Yes, I dare. I am.

Saturday, 2 July 2022 – Silverstone Circuit, Northamptonshire, UK

7 a.m.

I didn't sleep too well last night. I'm not sure why but for some reason it feels like there's a lot riding on this race. Gene hasn't been putting me under any pressure or anything, but I have this feeling. Sure, he's frustrated, but he knows that we're doing everything we can to turn things around. At the end of the day, though, Haas has scored two points from the last seven races. There's no getting away from that. As team principal I'm the one responsible, although I think the pressure I'm feeling at the moment is probably self-inflicted. I keep asking myself if I could do more to help the situation and I cannot think of anything. Is it just bad luck? I don't foking know. Because I'm the man in charge the team look to me when times are tough and to be honest I'm running out of things to say to them.

Look at this. I'm getting myself into a state. Yesterday was OK and as long as we're here we have a chance. That's right, isn't it?

10 p.m.

Kevin qualified seventeenth and Mick nineteenth. I don't know what to say. It feels like 2021 all over again. Everybody has been very down since then. From the guys in the hospitality unit to the guys in the garage. And I mean, really down. Week after week they hear talk about this great car we have and week after week we keep on failing to deliver. You really feel it in a small team. The highs are felt by everybody very keenly but, Jeezoz, so are the lows. There's no hiding place. I was going to try and speak to everybody to try and motivate them but they've heard it all before. There comes a time when you cannot motivate people just with words. The only things that will motivate this team of people at the moment are points and good results.

Sunday, 3 July 2022 – Silverstone, Northamptonshire, UK

7 p.m.

DOUBLE FOKING POINTS! After five races with fok all and then a crappy qualifying yesterday, we suddenly get double points. I'm still in a state of shock. And it's not like they were handed to us. We had to work our way into the points and fight for every single one of them. The car pace was good, the strategies were right and the drivers were great. Mick even had a good battle with Verstappen at the end of the race, which was fun. I did worry a little bit that he might do something stupid at the final corner but he didn't. He was sensible. At least Mick has some points now and he drove a very good race. I'm pleased for him.

The first person I called after the race was obviously Gene. Somebody once asked me how Gene Haas expresses joy and I'm still not able to answer that. He's a typical Californian really. He doesn't get very emotional and gives nothing away. That's not a criticism. He is who he is. In the beginning I had to try and work out what made Gene tick and that was one of the first things I noticed. When I told him about the result today the first thing he said was, 'So when are we getting a podium?' I said, 'Look, I'm just happy that we're out of the shit for the time being. One step at a time, Gene!' He wasn't joking, though. Gene Haas is a very focused and driven human being.

The big challenge now is obviously to see if we can maintain this form and score points next week in Austria. That's what Gene will be expecting and so will I. Also, it's another sprint race weekend so we'll have double the chances. Perhaps the worm is finally starting to turn? Oh my God, I hope so!

Thursday, 7 July 2022 – Team Hotel, Spielberg, Austria

9 p.m.

Despite it being a double header, everyone at Haas is full of energy still. Any points are obviously welcome but scoring double points gives you an extra lift. Both sides of the garage have something to celebrate so nobody is left out. Wouldn't it be great if we could replicate that here? Gene's flying over for the race so I hope it happens.

I did an interview on Teams earlier about sprint races. I know not everybody thinks this way but I really like the concept. Spectators want to watch competitive events so to have one on Friday, Saturday and Sunday can only be good. The feedback from last year's sprint races was positive and the feedback from the one we've had this year has been even more so. The only thing I would change would be to ditch FP2 on a Saturday. It's pointless. The engineers love it but at the end of the day we're not racing for them. We race for the spectators and they find it boring. Instead, why not do a second qualifying session for the Sunday? I've already spoken to Stefano about this and he agrees with me. And if it doesn't work, change it back again. Never be afraid to go back to something if you have to. The most important thing is that you give something a go. I think the highest number of sprint races we could do in one season would be eight or maybe ten.

No more, though. And you'd have to mix them up a bit, you know. Maybe use them as an incentive. If a promotor in one of the countries does a good job and brings in some additional sponsorship or something, reward them with a sprint race. It's another asset and one we have to use wisely.

Anyway, we've got qualifying tomorrow so I'm going to have an early night.

Friday, 8 July 2022 – Red Bull Ring, Spielberg, Austria

2 p.m.

I had to go to an F1 Commission meeting earlier and when I came out FP1 had already started. 'Anything to report, guys?' I asked. 'No,' they said. 'Everything's going to plan.' Everybody in the garage was very relaxed and both drivers were in the top ten. Lately we've really struggled in FP1 so I was expecting the atmosphere to be quite frenetic. I much prefer the new way!

6 p.m.

Qualifying was like an extended version of FP1 really. Everyone was very calm and relaxed and before I knew it both cars had reached Q3. Kevin ended up qualifying sixth and Mick seventh. I thought, *What the hell is happening here?* The last time we were in a situation like this was in 2019. It was almost too easy really. As I always say, though, being bad takes a lot more work than being good.

Saturday, 9 July 2022 – Red Bull Ring, Spielberg, Austria

5 p.m.

What an afternoon. And what a race! They certainly put on a good show for the fans. Because there was a danger that Mick and Kevin might be racing each other from the start, I sat them down before the race and made it clear that we would decide who is faster, not them. After all, we'd have all the data. If you have two cars from the same team racing each other with three DRS zones, the guy behind will always think he's faster and will be on the radio the whole time saying so. If the driver behind is only faster on the straight, there's no point letting them pass as all they'll do is lose momentum during the manoeuvre. If they're also quicker through the corners, however, it obviously makes sense.

What happened during the race was exactly that, except Mick, who was chasing Kevin, was being chased by Lewis Hamilton. In this situation we had to think about the team. Mick actually might have been a bit faster than Kevin but had we let him pass there was a danger that Lewis could have passed both drivers, which would have cost us points. It was hard for Mick to take but he had to hold his position and in the end Hamilton passed him with two laps to go. He didn't manage to pass Kevin, though, thanks partly to Mick, so Haas came away with two points. The most unfortunate thing for Mick is that, had we let him attempt to pass Kevin, he might have been the one who scored those points and instead he scored none. Then again, Lewis might have taken both of them and we'd have ended up with just one. At the end of the day, we all work for

Haas F1 and the team will always come first. I explained that to Mick afterwards and he was fine. He's a young driver who is hungry for success, though, and if he hadn't been upset I'd have been worried.

9 p.m.

It's safe to say that I am not a favourite person of the German F1 media tonight. Then again, am I ever? Mick could win the foking Championship and they'd complain that he didn't win by enough points. I get that they want him to do well but Mick Schumacher works for Haas, not the other way around. He's entitled to the team's support and he gets it but not at any cost.

What worries me a little bit is how influenced Mick might be by the German media. As you already know, there are one or two people there who have it in for me and if Mick believed everything they wrote and said he'd think I was wishing for him to fail. I'm obviously not but there are people who think I am. Because of who he is Mick is in a very difficult position and I don't envy him. The pressure to succeed when you're a Schumacher must be crippling sometimes. I will say this, though, if some elements of the German media didn't make up as much shit as they do, Mick and I would find it easier to develop a relationship, which would in turn help Mick to progress. They – not me – are holding him back.

Anyway, enough of that bullshit. Today we have two more points than we did yesterday and tomorrow we start seventh and ninth. If somebody had offered that to me when I arrived the other day I'd have bitten their hand off.

Forza Haas!

Sunday, 10 July 2022 – Red Bull Ring, Spielberg, Austria

4 p.m.

I kept a lookout for Sky Sports Germany after the race but I couldn't find them. And why couldn't I find them? Because poor badly done-to Mick who gets no support from Guenther finished sixth today and scored eight foking points! Put that in your pipe and smoke it.

In two race weekends we've gone from being a bunch of wankers in ninth in the Constructors' to a bunch of good guys in seventh. I actually saw Gene smile earlier. Seriously! I saw it with my own eyes. He then spoilt it all by asking me when we were going to be sixth in the Constructors'. I said, 'Give me a break, Gene!'

I'm starting to dream again now. At one point in 2018 Kevin scored points in five out of six races and Romain scored points in four of them. When something like that happens, all kinds of things start running through your head. You know, podiums, etc. It's almost as if a new part of your brain kicks into gear. *Shit, I remember these thoughts! Can I have more, please?* Despite all the dreams, what I really want is for us to be able to build on what we've achieved so far and not go backwards. If we can do that, I'll be a happy man.

Tuesday, 12 July 2022 – Castello Steiner, Northern Italy

10 a.m.

Every so often I will suddenly start receiving a load of text and WhatsApp messages from friends of mine saying things like, 'Fok me, look what they've done to you this time!' It always means the same thing, that our creative team have done something that makes me look like a wanker. It started again yesterday when I received a message from a friend of mine saying, 'Jeezoz Christ. Tom Cruise has let himself go!' I immediately thought to myself, *Oh fok. What have they done this time?* Over the next few minutes I received about ten more similar messages and then finally somebody sent me a link. I'm not sure how to explain what I saw really. There is a new *Top Gun* film out shortly and what the creative team have done is make a new version of the poster, just like they did with Grand Theft Auto for Miami. The photo of Tom Cruise has been replaced by me, looking as cool as fok in a pair of sunglasses, and they have changed the title from *Top Gun* to *Top Gunth*. I hope they put a photo of it in here somewhere because it looks quite funny. It's certainly gone down well online, apparently. Stuart made a point of sending me a response to it on Instagram that says, 'Haas admin deserve a pay rise.' What's next, though? *The Gunthfather*? *Raiders of the Lost Gunth*? *Poltergunth*? *Gunthfellas*? *La Dolce Guenther*? At least we know it has legs. I think we are one of the best teams when it comes to the online stuff. Our guys are funny and they know what they're doing. They can whistle for a pay rise, though.

Monday, 18 July 2022 – Steiner Ranch, North Carolina, USA

Well, what I call the silly season has finally kicked into gear. About this time every year the representatives of the drivers who are out of contract at the end of the season start contacting the team principals and the merry-go-round starts once again. Not to mention the representatives of all the drivers who are trying to get back into Formula 1 or break into it for the first time. That's been taking up most of my time for the past few days and already I'm fed up with it! It's a necessary evil, though, because as well as being prepared if a driver decides to leave, you have to keep your options open if you decide you want a change. Mick is out of contract at the end of the season and pretty soon his guys will be on to me about a new contract. Or not, as the case may be. Who knows, he could have other ideas. With Kevin we have a multi-year contract but that doesn't mean he's locked in. Again, he could decide to leave or we could decide that we want a change.

Mick has had one season as a rookie and half a season as what the Americans would call a second-year sophomore. Sure, he's shown promise, but until he's scoring points regularly and challenging Kevin I won't be happy. I think he believes that he's good enough to race for a championship-winning team but he's got a long way to go to achieve that in my opinion. There are twenty-two races this year and three or four good performances is not a ratio that any Formula 1 team is looking for, least of all a championship-winning team. Formula 1 is all about progression and what you cannot do as a team principal is allow anything to hold you back. It's the same for Kevin, by the way. Just because he's on a multi-year contract and has been doing quite well doesn't mean he is safe from any kind of criticism. He too needs to perform at every race and we're only halfway through the season. If I ignored silly season and didn't keep an eye on things I wouldn't be doing my job. It's as simple as that.

The other bit of news from Haas land is that we're making good progress with the new title sponsor. I can't say too much at the moment but hopefully we'll put pen to paper in the next few weeks. If it comes off they'll be a perfect fit for Haas, so fingers crossed.

Anyway, I've got a plane to catch. I've got a few days in the UK for more meetings about absolutely foking everything and then a Grand Prix to go to in France. *Au revoir, mes amis.*

Friday, 22 July 2022 – Circuit Paul Ricard, Le Castellet, France

10 a.m.

The weather here is amazing at the moment. Not a cloud in the sky anywhere. They're expecting the track temperature to hit 60 degrees later so I'm glad I'm not a driver! A lot of people have been asking me about Austria since I arrived. A strong performance in practice, Q3 for both cars in qualifying, points scored in the sprint, double points in the Grand Prix. I suppose it's a good story. Somebody asked me if it was a perfect weekend. 'Almost,' I think I said. It doesn't matter who you are, perfection is only achieved by winning. What I keep pointing out, though, is that Austria and Silverstone weren't some kind of miracle. They happened because we have a good car and because everybody did their job well. We're not punching above our weight here. We're where we are because we deserve it. The question now on everybody's lips — mine, Gene's, the team's and the media's — is can we maintain it. Jeezoz, I hope so. This whole success thing is pretty foking moreish!

4 p.m.

Free Practice was solid enough. Nothing major to report. No major issue, though, which is the main thing, and both drivers are happy. Hot, but happy! Just the start we wanted, really. Early night for me.

Saturday, 23 July 2022 – Circuit Paul Ricard, Le Castellet, France

5 p.m.

Jeezoz Christ. Lady Luck didn't just desert me so far this weekend. She shat in my sock drawer before she left! Wow, talk about frustrating. Kevin qualified eighth in Q2 but had to have an engine change, which means he exceeded elements with the power unit and will have to start from the back of the grid. Mick was unable to get out of Q1 after his fastest lap of the session was deleted for exceeding track limits at Turn 3, which means he'll start nineteenth. The ironic thing is that Kevin had already qualified for Q3 so didn't need to go back out, which is when we had to change the engine. It's nobody's fault, though. Just bad luck. On the plus side, the car definitely has enough speed to get us up the grid and, dare I say it, even score us some points? Lady Luck might have disappeared but my optimism hasn't.

Sunday, 24 July 2022 – Circuit Paul Ricard, Le Castellet, France

6 p.m.

Today was like yesterday. A shit afternoon but not without promise. Kevin was on fire at the start and went from twentieth on the grid to twelfth in no time at all. Then the safety car came out, which is when it all went to shit. Every car except Kevin's and Mick's was on a one-stop strategy so when the safety car went out they all got a free stop. Zhou then made contact with Mick, who spun and never really recovered, and Kevin collided with Latifi, which spelled the end of his race. On the plus side? The car once again showed that it's fast. We just need to regroup now and hope that Lady Luck comes back to me soon.

Thursday, 28 July 2022 – Team Hotel, Mogyoród, Hungary

I'm not sure if there are any rumours circulating but once again I've been getting a lot of love from drivers and their representatives this week. Funnily enough, Gene and I have actually started talking about the driver situation for next year now so perhaps they're all psychic? We cannot make a decision yet, though. We want to give Mick as many opportunities as possible to prove himself and at the moment the jury is still out. We also need to know who will be available for next season. It's a short list just now. The other reason is politics. You have to be careful not to get caught up in any ongoing battles so it's better to keep quiet for as long as possible.

I think I'm allowed to say now that we have a 'bookies' favourite' for the title sponsorship. Until yesterday there were still two contenders, but the one who appears to be most keen came back to us today and improved their offer. No offence to the other contender, but these guys are an American company (the one I mentioned last time) and they're seriously keen to work with us. This, almost as much as what we've achieved with the car this season, validates the decisions we've made. We're almost at the heads of terms stage, so it's pretty advanced now.

Friday, 29 July 2022 – Hungaroring, Mogyoród, Hungary

5 p.m.

I have the biggest news in Formula 1 this week. Bigger than anything at Mercedes or Red Bull. Are you ready for this?

Haas has an upgrade!!

Formula 1 can rest easy now. The upgrade is here.

It was originally intended for France but it got pushed back by a week. Only Kevin's car will have it here but if the data and the numbers match the wind tunnel then both cars will have it after the summer break.

Anyway, after practice today it's clear that we've still got some work to do as there were some balance issues during practice. Nothing massive changed, though, and Kevin says that the car feels more or less the same in terms of characteristics. You see, this is not a magic bullet. It's taken us a long time to get to know the car as well as we do and it will be a similar process with the upgrade. A shorter process, hopefully.

Saturday, 30 July 2022 – Hungaroring, Mogyoród, Hungary

10 a.m.

The feedback to the upgrade from outside Haas has been entertaining. We've been accused of copying Ferrari. In fact, according to Stuart, some people online are now calling our car a white Ferrari, which is imaginative. Look, at the end of the day there are three concepts out there: the Ferrari concept, the Red Bull concept and the Mercedes concept. We are close to Ferrari, so obviously we're going to copy that one. We have the same engine, same gearbox, the same suspension. Come on, why would we copy anything else? And they're winning races. Don't worry, I can take the criticism. Our cars have been called worse things and so have I. Sometimes fairly! It'll be interesting to see what happens in qualifying. We're managing our expectations, though. As I said yesterday, upgrades are not a magic bullet.

4 p.m.

Qualifying went OK. We got out of Q1 pretty well, but we just didn't have the pace to get from Q2 into Q3. Kevin's P13 and Mick P15. Anyway, we're focused now on tomorrow. If we get a good start there's no reason why we can't finish in the points. We're here to fight and tomorrow that's exactly what we'll be doing.

Sunday, 31 July 2022 – Hungaroring, Mogyoród, Hungary

9 p.m.

If I said what I really think of the FIA at this very moment in time I'd receive a lifetime foking ban! OK, where do I begin? Well, at the start of the race, Kevin's front wing sustained some damage after a small collision with Ricciardo. And I mean small. Nothing serious at all and certainly nothing dangerous. Ocon then gets on the radio and starts complaining about it and before we know it Kevin has been given a black and orange flag.

It had already been agreed with the FIA that in situations like this they would call the team first and ask for their opinion. After all, cars sustain damage all the time during races and the people who are best placed to know the extent of that damage are the teams. Had they called us like they'd agreed to we would have told them that the damage to Kevin's wing was negligible and that it was safe for him to carry on. Not just safe for him, but safe for everybody. Had they disagreed with that, we would have had a discussion about it and we would have decided that I was right. The point being that instead of calling us and asking for our opinion like we'd agreed, they did Ocon's bidding and just showed Kevin the foking flag.

The rules state that when a driver receives a black and orange flag they have to come in the next time they pass the pits, and when Kevin re-joined the race after we changed the wing he was right at the back of the field. Had we delayed the stop and tried to contact race control, Kevin would have been disqualified so we had no choice. We had to wait until afterwards.

When I called race control after the race they denied having any knowledge of the agreement. 'But I've got the foking letter!' I said to them. 'The letter that you signed!' I'm sending them a copy tomorrow. The guy then tried to tell me that he thought the wing was going to fall off, which genuinely is just bullshit. We know how it's designed and the reason I know it wouldn't have fallen off is because there is Zylon on top of it, and Zylon doesn't break. We obviously would have changed the wing because we were losing downforce, but when the time was right for us. Not for Ocon! Even a lap later would have saved us at least ten seconds because it was under the virtual safety car. Kevin was going well in P13 when this happened so had a very good chance of getting in the points. Lady Luck is obviously having an affair with these people.

We already have some history with this. In Canada, when Kevin made contact with Hamilton, he was issued with a black and orange flag for what was basically a scratch on his front wing. Just like this time, he was forced to pit early, which screwed up our strategy. Last year in Jeddah, Hamilton won the foking race with only half a front wing! Where's the consistency? You know, I'm really disappointed with this. It has to change.

Anyway, what else can I tell you? We had a nightmare with the tyres this race. Kevin changed on to a hard compound when he pitted for the front wing and because of the blue flags that followed he could never get them working. Mick had a similar problem. He finished P14 and Kevin P16. It's not the way I wanted to go into the summer break but what can you do? You can't legislate for shit like this. We have a month off now. A month off to regroup and come back fighting.

Now, where have I heard that before?

SUMMER BREAK

Saturday, 6 August 2022 – Somewhere nice in Tuscany

Quick update from Guenther land. I've been sunning myself in Tuscany with Gertie and Greta. It's not very often I admit this but I was pretty tired by the end of Hungary and I was ready for a break. It's not just the physical part. The mental part is just as consuming and a few days without any distractions has been just what the psychiatric doctor ordered. Anyway, I'll be back to the grind in the next few days so we'll catch up then.

Monday, 15 August 2022 – Castello Steiner, Northern Italy

As much as I like being on holiday with my family, I wish the summer break was not this long. You can have too much of a good thing, you know. I understand why they do it but it's not for me. Luckily, late last week Stuart called me up and asked me to do an interview on Wednesday, which I might need to prepare a bit for. Or at least think about. I am never doing that usually for interviews so it will make a change. Apparently they want to talk about lots of things including infrastructure, sustainability, competition, funding and inclusivity. I said to Stuart, 'What do they foking want, a state of the union address?' 'Basically, yes,' he replied. So I thought, *While I'm thinking about what to say, why don't I put it in here?* You cannot tell me that you are not dying to hear what I have to say about things? The world according to Guenther Steiner? Well, tough shit, because that is what I am going to do. Come on, we still have almost two weeks until the next race. Work with me here! It'll help to fill in the time.

Because of the way Formula 1 has grown over the last few years, and because of the way it now engages with fans and with the media, questions and conversations about things like the future of the sport, it's structure, sustainability, funding, competition, regulations and environmental pressures have become an everyday thing. Personally, I'm more than OK with that because these are conversations that we should be having regularly. They have also helped the sport develop a conscience, which I don't think it really had before. Not to mention a sense of self-awareness. You no longer need to shine a light on what F1 is doing because it is doing that itself. Sure, it's not perfect, but things like sustainability and taking care of the environment have now become part of the sport's culture, whereas before I think it was just tokenism. Ticking a box, in other words. Do it because you have to. As a major global sport, we have a responsibility to lead the way with issues like these and generally I think we're doing a pretty good job.

What's that you say? Back it up, Guenther, you idiot. OK, I will try.

With regards to the future of the sport, I would say that Formula 1 has put itself in a good position in the medium to long term just by engaging with younger fans. As long as they carry on doing that and don't forget about what they are doing, it should be OK. I don't know the exact figures but the difference between an average F1 racegoer in 2001, which is when I joined the sport, compared to 2022 is pretty noticeable. For a start, there are a lot more of them! That's for sure. But they are also a lot younger these days and there are also far more women at races than there used to be. That's all positive.

The one thing I think Formula 1 needs to do in order to secure its mid- to long-term future is stabilize and refine what has already been achieved. The growth it has experienced over the last few years has been crazy and I'm sure it must be tempting for the people at the very top of the sport to let it carry on. After all, Formula 1 is a commercial enterprise and when a commercial enterprise shows this kind of growth you don't want it to stop. The sport itself needs to keep up with that growth and it can only do that by taking a step back, evaluating what has happened and making little changes where necessary. That – the refining part – is something that Formula 1 has become very good at over recent years. After all, taking care of the fine detail has always been a fundamental part of what we do, so why not do it with everything?

The good thing is that the dialogue between the teams, Formula 1, the FIA and everyone else who has a stake in the sport is ongoing, so as long as that continues I think we should be fine. There's no excuse these days for not communicating with people. Not like in the olden days when nobody had a cell phone or a computer even. Everyone could just ignore everyone else. And that's what we did, which meant you always had an excuse for not doing things. Having a cell phone can become a little bit intrusive sometimes. I get that. There are a lot of positives to having one, though, and one of them – the main one, I'd say – is being able to convey a message to one person or even several million people in a few seconds. With regards to the communication within Haas, there are three steps: we discuss it, we decide that I am right, and then we carry on.

Funding is a subject I get asked about a lot these days, and especially since the shit show with Uralkali. Funding has always been a hot topic within F1 and that's obviously because of the money involved in the sport. The downside to everyone having cell phones and computers is that a lot of false information gets thrown around and that can lead to problems. One of the funniest things I read on the internet is what the drivers are supposed to get paid. A lot of that is just guesswork, as is a lot of the other bullshit I read. The upside is that there is now a lot of very good content available out there regarding F1, so it's swings and roundabouts. Look at the man who takes the piss out of me on Twitter. Can you imagine a world without him? I cannot.

With regards to funding, though, that is in a very good place. There is a very high demand for Formula 1 at the moment, and where there is demand, there is money. The interest we as a team have received this season, not just from potential sponsors but from people who want to invest in the sport, has been incredible. I cannot speak for F1 as a whole, of course, but if we are getting a lot of interest then you can probably guess that the rest are too.

Funding within Formula 1 is also in a good spot at the moment, primarily because of the budget cut. That has also been reached in just three years (since the last commercial agreement), which has been a hell of an achievement. The bigger teams obviously didn't like it, but what can you do? If an advantage gets taken away from a team, they're bound to make a fuss. Not only because of the advantage being taken away, but also because of the restructuring they have to do. They just need to be better, that's all.

Before, the team with the biggest budget would almost always win the championships. Now, the focus is on talent. Talent, and managing the money you have effectively. You remember what I said before the Monaco Grand Prix about Christian complaining about expenditure? It turns out that the reason they wanted the budget to be increased was because they had gone over it. Damn these level playing fields. They can cause foking havoc sometimes!

I suppose this fits in funding, but a question I am asked a lot these days is whether or not we should have more teams in the paddock. The reason I am getting asked this question a lot at the moment is because Michael Andretti has been talking for a long time about entering a team and it's fair to say that some of his comments lately have not been very constructive. I don't know. In my experience, if you want to invest in an organization that is already successful, as well as telling the people at the top what you can bring to the party, you have to try and make a good relationship with everyone. That would seem like the most sensible way to do it. Even if your investment isn't accepted immediately, the best way forward would surely be to work out why it hasn't been accepted and then come back with a better offer. Formula 1 was in a good place when Michael started talking about this, but it's in an even better place now. It's a sellers' market, in other words.

From my own point of view, if somebody can prove that they can help to increase the revenue of our sport by at least 5 per cent, then why not? But if somebody just wants to come in because we are doing well, then no, fok off. You have to bring something to the table other than just your presence. Formula 1 is obviously very aware of that, which is probably why the applications that have been submitted recently have not been successful.

When Gene Haas took a chance to start a team with me back in 2014, Formula 1 was in a very different place. It wasn't in a bad place, but it was nowhere near as popular as it is now. Over the years he has invested many millions of dollars into the team and the sport and sometimes during some pretty uncertain times. Sorry, some very uncertain times! Why would he want to have that investment diluted by having another team on the grid if they weren't going to improve things? In 2015 you could buy the Manor F1 team for one pound, but nobody wanted to. Now, the starting price for an existing F1 team must be about half a billion dollars. No wonder everybody wants to join. But where were they seven years ago?

Although I'm not against a new team coming into F1 – or new teams, for that matter – I think the way things are now is pretty good. When was the last time Formula 1 had ten stable teams on the grid? I can't remember it ever being like this. The best way to get into Formula 1 at the moment, then, would be to buy one of the existing teams. Everything is for sale. That, I do know. It just depends on the price.

Anyway, that's enough for one day, I think.

Tuesday, 16 August 2022 – Castello Steiner, Northern Italy

Ever since the team started, the relationship I have had with our sponsors has generally been very good. Like everything, though, the more you learn about the people and the company that is sponsoring you, the more things can improve. Quite a few of our sponsors ask me to go into their offices and talk shit sometimes, and I'm always happy to do it. Why they ask me I just don't know, but they keep on doing it. I never make notes or put a script together. Just like I did when Gene and I applied for the licence, I just make some bullet points, go in, and start talking. I don't always stop talking, but at least I do it.

Just like Formula 1, my 'flying by the seat of your pants' approach to making presentations and speeches has moved on a little bit and I've even had to make some changes.

Let me give you a little example.

A while ago now one of our sponsors asked me if I would go in and do a forty-five-minute talk to their staff. 'Sure!' I said. 'I'd be delighted.' They then sent me four points that they wanted me to cover in the talk, which I forgot to read, and so I was good to go. The following day, about an hour before I set off for the talk, I received an email from them asking me for my slides.

'My what?' I said.

'Your slides, Guenther. For your talk. On sustainability?'

'On what? Oh shit!'

When people ask me to speak for forty-five minutes I usually offer them an hour and forty-five, on the understanding that they just let me talk my usual bullshit. That's what I excel at and if you gave me enough notice I could probably do my own festival. Guentherstock! What I'm not very good at is being given specific subjects. Well, I am, but only when I remember to read the foking email telling me about it.

I rang up the sponsor and said, 'Guys, I think that we have a problem here. I'm afraid I have failed to prepare, which means I have prepared to fail!' 'It's OK,' they said. 'We'll get somebody to ask you questions and turn it into a discussion.'

They created a monster by doing that because there is only one situation where I can speak more than if I am on my own, and that is when I am with somebody who is feeding me questions. Luckily, the person asking the questions knew when to stop and so they escaped with about an hour. Lucky fokers!

Anyway, that little story leads me nicely into the subject of sustainability. Which, since that talk, I have become an expert in. To be serious, it is something that we all have to think about these days. As with things like inclusion, sustainability and the environment have become part of our everyday thinking so it's always a consideration.

What I would say first on the matter is that it starts with the individual. Since becoming more aware about sustainability and the environment, I have become more aware of my own actions and I keep on trying to get better. You can't do any more than that really. And this is not some corporate bullshit line that has been fed to me by Stuart. This is actually what is happening right now. Having a teenage daughter helps. To young people things like this are second nature these days and fortunately they lead from the front.

Formula 1 too needs to lead by example and at Haas we take the environment and sustainability very, very seriously. We have no foking choice now that I am an expert! We are working towards our Three Star Environmental Accreditation by the FIA, which is recognition for the fact that sustainability and the environment are now a constant consideration in everything we do. Every team is the same now, I think. And even every circuit.

Take freight, for example. A while ago now we started looking at different methods of transporting our gear around the world and very quickly realized that instead of just piling everything on to some cargo planes, we should look at sending it by sea. It turned out that not only is sea freight more environmentally friendly than air freight, but it's cheaper, too, and investing less at the front end means we can spend more at the back end.

The idea is for the sport to have a net zero carbon footprint by 2030. It isn't going to be easy but every team and every circuit on the calendar is working for that exact same goal and so if one of them isn't pulling their weight, everyone will know about it. That's what I mean about making it part of our culture. Also, you can bet your bottom dollar there will be some competition there. Mercedes will not want to be outdone by Red Bull and Ferrari, and vice versa. It's a perfect situation really. Certainly in terms of setting an example. Think about it. By asking ten Formula 1 teams to achieve a zero net carbon footprint you're basically starting a race, and one that any one of the teams could win. Whichever team does win will have the bragging rights and will get a lot of very good PR from it. And there's nothing wrong with that, by the way.

The net zero initiative also includes changes to the cars. By 2026, 50 per cent of the power units will be powered by combustion engine and 50 per cent by electricity. The cars will have less downforce but everyone will be in the same boat. Sometimes to do the right thing you have to adapt and Formula 1 is very good at that. Certainly from a technology point of view.

It will be interesting to see how this affects Formula E. Surely, the closer Formula 1 gets to becoming carbon neutral the less need there will be for something like Formula E? I could be wrong, of course. I think that Formula E has done a good job in making people aware of the challenges that are facing us. Does it have a future, though? Formula 1 is at a level where it is almost untouchable. It's such a big sport. Also, we have no idea yet whether electricity will be the future. It will certainly be part of the future, but there are many technologies being developed at the moment (many of them within Formula 1) and so putting all of your eggs into one technological basket is not sensible, in my opinion. If I had to have a guess right now I would say that the future of motorsport is not Formula E. There's also a lot of coming and going in Formula E; most notably, Audi is leaving Formula E to come to Formula 1. What does that tell you? Lastly, and I am sorry to stick the knife in like this, but the cars sound like foking sewing machines and the tyres make noises like sneakers do in basketball games, which gets right on my foking nerves! Knowing my luck I'll be completely wrong about this and in ten years' time I'll be out of a job and pretending I never said anything.

Including the circuits in the carbon neutral initiative has been an important move, I think. Sure, the teams are there for the whole year, but think about what is involved in setting up a Grand Prix and then having three or four hundred thousand people descend on your circuit for the weekend. Think about the logistics involved. Not to mention things like recycling.

The motivation behind everything I've just mentioned, in addition to cutting our emissions, recycling and reducing waste, etc., is setting the right example. The idea being from a spectator's point of view, if F1 are doing it, I should be doing it. Even old idiots like me can become aware of this stuff. These days I try not to buy water in plastic bottles and when I go to Walmart in my shorts and sandals looking dishevelled I make sure I don't use any plastic bags. The reason I've become aware of it is mainly down to what's been happening in our sport. It isn't up for discussion any more. It's part of who we are and what we do.

The next thing I want to talk about is inclusivity, which is another important subject right now. Earlier, I said that the budget cut will force us to focus on talent more instead of just money. That should be across the board, I think. Formula 1 has a reputation for being quite elitist and that is something that has to change, in my opinion. Not just from an inclusivity point of view, but from a talent point of view as well. After all, the more people who take part in entry-level motorsport, the more chance we will have of finding the next Lewis Hamilton or Max Verstappen. It's a dilemma, though. For a start, motorsport at any level is an expensive business and there's no escaping that. It's not like soccer where all you need is a ball. Also, because the sport is so popular right now, there could be a danger of overcrowding. As usual, it's all about finding a balance. The public want to take part, we want to discover the talent, but who is going to pay? Same old story. It's not going to happen overnight but initiatives are being put in place globally to make things like karting more accessible to people. I'm actually quite optimistic about this and although we have to make things more open I believe that if somebody has the talent and the drive, they will make it. Maybe I'm being naive about that but it's what I think.

The most common question I am asked with regards to inclusivity is why do we not have more people of colour and more women working in Formula 1. My answer to that at the moment is that the sport hasn't focused on developing that talent, which is true. As with young drivers, Formula 1 (and motorsport in general, I think) has to, and is doing, more to change people's perception about who can work in motorsport. And, more importantly, who we want to *have* working in motorsport. As soon as women or people of colour are working here nobody looks at them differently. As I said, it is all about perception. It's motorsport's problem to solve, though. Make no mistake about that.

We as a team and me as a human being have never given or refused to give a job to somebody because of the colour of their skin or their gender. I don't care who you are. If you're good enough and there is a position available in our team, you will always stand a chance with me and I do not know anybody in our sport who thinks differently. Formula 1 as a sport, and as an industry, always wants the best people available. That is where it starts and that is where it stops with us. The question is, though, what can we do about it? Well, apart from altering people's perception, we need to focus on recruiting from underrepresented groups. Just like sustainability and the environment, it has now become part of our culture and I'm confident that things will improve over time.

Stuart is going to have a foking fit when he reads this next bit, but in 2023 there will be a Grand Prix in Qatar as there was in 2021 and I have an opinion on that. I promise you, he will be sweating like a pig in a sausage factory at the moment! My opinion was actually formed when I first found out that we were going to be racing in Saudi Arabia. My opinion then was based on what I'd heard and read in the media so I decided to do my own research. I ended up speaking to some people I knew from Saudi Arabia and from other parts of Arabia and they all said the same thing, that events like Formula 1 will help places like Saudi Arabia and Qatar to progress and to make important changes. They also said that the problem they have is with the older generations and so things cannot change just like that. The point is, though, that according to them they *can* change and they will, which is one of the reasons why I believe that we should try and lead by example and race in these countries. And I don't mean that we try to force them to immediately adopt Western values. Who wants that? We're not perfect, you know. They learn from us and we learn from them. And if we don't agree with some of their beliefs and laws, try and do something to change them. Formula 1 is a global sport and we should be careful about choosing to ignore countries because of certain beliefs. Change nothing and nothing will change.

I had a discussion about this with a friend of mine a few weeks ago and he made the point about homosexuality being illegal in Qatar. 'OK,' I said. 'But will it always be illegal there? Also, might a pro-inclusivity sport like Formula 1 visiting the country help to change things?' I'll tell you one thing, it certainly won't do any harm. And let's not forget about some of the attitudes towards LGBTQ+ people in the places where we are already racing. Look at Hungary, for instance. They're part of NATO and part of the European Union, yet same-sex couples are not eligible for some of the same legal rights as heterosexual couples. Things are not perfect there but they have been improving. Sure, we are not responsible for that improvement, but I'd sooner be with the people of Hungary, Qatar and Saudi Arabia as they slowly move away from these kinds of prejudices than just stand on the sidelines and disapprove of them. A friend of mine is the former F1 journalist and PR chief Matt Bishop. He is a gay man and he is of the opinion that it is better to go to these places and be part of the change than to boycott them. It's a brave thing for him to do, you know. Especially in a country where it is illegal to be a homosexual. It is the right thing to do, though, and I, and I'm sure everyone in Formula 1, stands with Matt and the LGBTQ+ community.

SEASON RESUMES

Friday, 26 August 2022 – Circuit de Spa-Francorchamps, Belgium

11 a.m.

I'm afraid the Belgian Tourist Board won't be as happy with me as the Azerbaijan Tourist Board was because, to be honest with you, I really do not want to be here right now. Spa is a fast track where you need low drag and unfortunately our car is not suitable. It is aerodynamically inefficient and we just won't have the pace. You never know, though. Anything can happen in this sport.

6 p.m.

Well, that was fun. Not! Mick will be starting the race from the back of the grid due to a grid penalty for exceeding the allowed number of power unit and component elements. Despite this, he remained positive and professional throughout the session so I couldn't have asked more from him. Kevin suffered a high-voltage system issue, which forced him to park at the exit of La Source before a red flag allowed them to recover the car. FP2 went without incident and he finished the session P17.

Right, I'm going for a glass of wine. It might be a pint glass the way today is going.

Saturday, 27 August 2022 – Circuit de Spa-Francorchamps, Belgium

6 p.m.

Jeezoz Christ. We actually had some luck today! Kevin qualified eighteenth in Q1 but because of some other power-unit-related grid penalties that were applied after qualifying he'll start the race in twelfth. Mick actually made it into Q2 and had he too not had the penalty he would be starting fifteenth. Just to bring me back to reality for a second, Kevin won't have the pace to maintain his position so, barring a miracle, he'll finish way further down. If I was a betting man, I'd guess that Kevin will finish in sixteenth and Mick eighteenth. I'm an optimist but I'm also a realist. It's just not a good track for us.

Sunday, 28 August 2022 – Circuit de Spa-Francorchamps, Belgium

7 p.m.

I was one out! Kevin finished sixteenth, as Mystic Guenther predicted, but Mick was seventeenth. Under the circumstances Mick did well. I'm still disappointed with the result but only because we haven't been competitive. It reminds me of last season. On a positive note, the guys back in the UK are already working on a package for here that will hopefully make us competitive next year. Anyway, we'll try to make it up next week in Zandvoort. Our car should be better suited to that track and, with some luck, it could end up being another Austria or Silverstone. I'm glad we only have to wait a week, though. I need to exorcize this race from my mind!

Tuesday, 30 August 2022 – Castello Steiner, Northern Italy

8 p.m.

Today has been one of the most difficult days I have had in over a decade. I hope you will feel sorry for me when I tell you. Basically, Netflix turned up at shit o'clock in the morning and took me to Mattia's place. He's just bought a vineyard and they thought it would be a good idea for him to show me around and then for us to taste wine together while pretending we know what we're talking about, eat food and then sit down and talk bollocks. I said, 'You're in luck. We're good at all of those!' When it's ready I'm going to buy a case of Mattia's wine, stick a Haas badge on it and then send it over to Andreas Seidl at McLaren with a card saying, 'Made from sour grapes – your favourite!'

After eating, drinking and talking bollocks at Mattia's place, we went to a winery near my place and did exactly the same again!

What a day, though. You must be crying your eyes out by now! Pray for Guenther!

Thursday, 1 September 2022 – Circuit Zandvoort, Zandvoort, Netherlands

2 p.m.

This is what we want! Plenty of medium- to high-speed corners, not many low-speed corners, and no ridiculously long straights. Nothing is guaranteed but at least we know that we stand a chance of being competitive. This race could actually be the middle of a shit sandwich for us as the race after this is Monza, which could be just as bad as Spa. Anyway, let's see.

What they've actually done to Zandvoort over the last few years is incredible. Take my word for it, it's a foking transformation! Although pretty legendary, Zandvoort had developed a reputation for being an old-fashioned circuit that you probably wouldn't visit unless you had to. Back in the 1970s and 1980s it was amazing. I remember watching it on TV when I was a kid. You had to drive along a beach to get to it, for fok's sake! I remember thinking, *Wow, how cool is that?* That's where it stayed, though – the 1970s and 1980s – and the last Grand Prix to take place here before last year was in 1985.

Instead of just creating a completely new circuit, they've modernized what was already there and by doing so they've managed to retain a lot of its original character. It's also very sustainable, certainly with regards to transportation. These days, instead of driving a car along the beach to get to Zandvoort, you either cycle or take a train from Amsterdam and then walk. I can't think of any other circuit that operates like that. What's not to like? I think the attendance will be about 350,000 this weekend and the only reason it isn't higher is because they don't have the space.

There is often a lot of resistance to modernizing classic circuits like Zandvoort and one of the reasons is because it's assumed that the track will have to change, too. As far as I know, the only thing they have done at Zandvoort to bring it up to F1 standards is to add some banking here and there, but apart from that it's the same track they've been using since 1999. Jeezoz, I'm beginning to sound like a nerd! I'm going to do some work now.

Friday, 2 September 2022 – Circuit Zandvoort, Zandvoort, Netherlands

10 a.m.

I went out for a meal last night and I was approached by two American guys from New York. 'Hey, Guenther,' they said. 'Could we get a selfie with you?' 'Sure, guys,' I said. 'I take it you are here for the Grand Prix?' 'Fok, no,' they said. 'We're here for the sex and drugs!' At least they were honest.

5 p.m.

Not a bad day on the track. FP1 was productive for both drivers, with lots of laps. It was especially good for Kevin as this is his first time at Zandvoort in an F1 car. We didn't quite replicate that form in FP2 but we got to complete our run through of all the tyre compounds. There's work to do, as always, but it's been a good start to the weekend.

OK, Mattia and I are going to find those two American guys!

Saturday, 3 September 2022 – Circuit Zandvoort, Zandvoort, Netherlands

10 a.m.

Jeezoz Christ. I can't make up my mind which fans are the craziest, the Australians or the Dutch. Given the noise they were making when I left the circuit last night, and the number of zombies that were dragging themselves around the circuit this morning, I'd have to say the Dutch. One guy was lying on the grass outside the paddock flat on his face earlier. Well, at least they're all enjoying themselves.

Drive to Survive must be pretty popular over here because I've been getting a lot of comments. Actually, what am I talking about? The world champion is foking Dutch! A group of guys yesterday shouted to me, 'Hey, Guenther, we look like a bunch of wankers!' I said, 'Too right you foking do!' It's a really cool atmosphere this weekend.

In some countries I get hardly any comments or selfie requests and, to be honest with you, it's quite nice sometimes. In Brazil, for instance, most people have no foking idea who I am so generally I'm left alone. I get one or two requests, but it's nothing like here or Australia. OK, see you after qualifying. I've got a good feeling about this.

5 p.m.

Well, it's been Mick's weekend so far. No doubt about it. Kevin had problems getting grip for some reason and wasn't happy with how the car felt. It's something we have to look at. Not so Mick. He got out of Q1 pretty easily but, to be honest, I thought that's as far as he would go. He ended up finishing P9 in Q2 so proved me wrong. He even made up another position in Q3 (thanks to a yellow flag) so starts eighth. I think he has a good chance of finishing in the points tomorrow.

Sunday, 4 September 2022 – Circuit Zandvoort, Zandvoort, Netherlands

6 p.m.

Well, it looks like Lady Luck was out getting drunk with the Dutch fans last night because she sure as hell wasn't in the Haas garage today. Mick came in on lap thirteen for a set of medium tyres and for some reason the front jack wouldn't release. This cost Mick at least ten seconds and by the time he re-joined the race he was way off the points. It was nobody's fault. It was just a mechanical failure. We must have practised pit stops at least a hundred times over the weekend and every time it was fine. Then the race comes and what happens? It foks up. What's even more frustrating is that we haven't been able to find out what was wrong with it because it's working fine again. It was literally just that one time. The time when we needed it to work! Things like this play with your mind a bit. Why did it have to happen then?

I feel a bit sorry for Mick at the moment. He's driven well this weekend and he hasn't done his chances of being offered another contract any harm at all. I'd say it's fifty–fifty at the moment. His driving has definitely improved and if he can keep up this performance level until the end of the season (and without writing off any more cars) he could be here next year. That's if he wants to stay with us. He too will have other options, so nothing is guaranteed.

My biggest doubt about Mick at the moment is whether or not he is a team player. That's always been the case right from the start. He says all the right things but does he really mean them? He also tries to play mind games with me sometimes, which is quite funny. I get this from a lot of quarters, not just him, and it's because I like to have a laugh and a joke. Good old Guenther. I can get one over on him easy enough. What does he know? People who know me well know that it's just an act and that occasionally I can even use it to my advantage. Underestimate Guenther Steiner at your peril.

Right, next stop Monza. It looks like the sandwich I mentioned earlier could now end up being three pieces of shit. Let's hope not.

Thursday, 8 September 2022 – Autodromo Nazionale Monza, Monza, Italy

12 p.m.

When I collected my luggage at the airport earlier, one of the wheels had come off my suitcase. When I saw it I thought, *Jeezoz Christ. I hope that's not an omen!* It wouldn't surprise me if it was. I hope nobody from the FIA saw me. If they did they'll probably give me a foking penalty. 'Hey, Guenther. We saw you with only one wheel on your suitcase. We've got something for you. A lovely black and orange flag!' Can they black-and-orange flag me for taking the piss out of them? Probably. Seriously, though, the last thing I need is a run-in with the FIA this weekend. I've got enough shit to worry about.

Friday, 9 September 2022 – Autodromo Nazionale Monza, Monza, Italy

11 a.m.

There are a shitload of people here this weekend. Stefano thinks they could get up to 350,000. That's a crazy number. Then again, we're in Italy and we Italians are crazy people. Despite my name I am actually an Italian citizen. I grew up on the Italian/Austrian border so I speak German and Italian fluently. I can swear better in English, but then you already know that. I had to do national service in Italy (every male over eighteen had to do a year until 2005) and it's safe to say that my surname didn't do me any favours. It was OK if your name was Pasquale or Rossi, but Steiner? I wasn't bullied but everyone thought I was a bit weird. Me, weird?

Do you know what I did during my year of national service? Absolutely fok all. What's more, I was bloody good at it. One day, early on in my year, the colonel's driver fell ill and I was asked to fill in. A few days later I got a call saying that he'd sacked his driver and wanted me to do it full time. So that's what I did. I drove an Italian colonel around sometimes and the rest of the time just scratched my backside.

Right, I'd better get my shit together. FP1 is coming up soon. This is where the fun begins. Or stops. I'm not sure which.

4 p.m.

Actually, that wasn't as bad as I was anticipating. Antonio Giovinazzi sat in for Mick in FP1 and did a good job, considering he hasn't been in a Formula 1 car for almost a year. The reason we put Antonio in for FP1, in layman's terms, was to help us see how good the car really is. If a driver who's never driven a car before can get in and get close to a driver who has, that's a good sign. He ended up being just three-tenths off Kevin's time, which was really encouraging. He did a good job.

When Mick took over in FP2, he had an electronic issue which meant he only managed to do three fast laps. That isn't ideal but we've got one more session with him tomorrow. Both of Kevin's sessions went without a hitch and he put in a lot of laps. Two out of three, then. Not bad.

Saturday, 10 September 2022 – Autodromo Nazionale Monza, Monza, Italy

8 a.m.

Silly season is still ongoing. Gene and I are talking every other day now and at some point we'll obviously have to make a decision. I'm going to make a stopover in LA on my way to Singapore for a meeting about it. Well, about that and everything else. I haven't seen Gene since Austria and we need to catch up properly. Drivers will be top of the agenda. Or at least close to it. The good thing is that there is no hurry to make a decision. There's a lot to consider. Both drivers have to fit in with our plans and ambitions for the future and the deciding factor is whether we think they can help us achieve them or not. It's not just about promise and talent. It's about ability and experience. Last year, having two rookies wasn't a problem as we weren't competitive. Now that we are again (and hopefully will be next year and in the future), we have to make sure that it isn't just the car that has become competitive. I think I said earlier that one of the biggest fears we have is not being able to realize the full potential of the car. Nothing has changed.

1.30 p.m.

Mick didn't manage to get out in FP3 due to a clutch issue, which means he's had very little preparation. It's nobody's fault, of course, but that doesn't make it any less frustrating.

I'm trying to think of something to write that will make this part of the book feel a bit less predictable and boring. OK, I'll tell you what was the last thing to make me piss my shorts. There's an account on Twitter called @BanterSteiner, which is basically a parody of me. Stuart made me aware of it a few years ago and every so often I have a look to see what he's saying about me. I've actually met him once. He came to see me at a race and asked if I minded him doing a parody. 'Go for your foking life,' I said. 'As long as you don't offend anyone, I don't mind.' Some of the things he comes up with are foking hilarious. And they're clever, too. The one I saw earlier today had a photo of Otmar at the press conference at the Dutch Grand Prix with the words: 'Here's my colleague Otmar during the press conference this morning warming up the chair I'm sitting in later. I swear his butt creates more heat than a foking tyre warmer!' I'm not sure what Otmar thinks about it but it made me laugh. And yes, the chair was quite warm when I sat in it. Thanks, Otmar.

6 p.m.

Kevin and Mick qualified nineteenth and twentieth but, because of some technical grid penalties that were applied afterwards, Kevin will start in sixteenth and Mick seventeenth. OK, so it's not quite as bad as we expected. Even so, Kevin's two fastest Q1 lap times were deleted for exceeding track limits and had that not happened he would have been through to Q2. And who knows, if Mick had been able to do some more laps he could have been there too. It's still shit, but it could have been shitter. There are also a lot of drivers out of position tomorrow, so you never know. We might even get a sniff at a few points.

Sunday, 11 September 2022 – Autodromo Nazionale Monza, Monza, Italy

11 a.m.

This has been the busiest Grand Prix of the season so far. There might not be as many people here as at Silverstone or Zandvoort but it's a much smaller circuit. It's been difficult to move around sometimes, especially when you have people asking for selfies. I'm not complaining. OK, maybe I am a little bit. It's just frustrating sometimes. At the end of the day I'm here to do a job and it's taking me ten or fifteen minutes just to walk the 4 or 5 metres from the hospitality unit to the garage. I know I should learn to say no sometimes but I find it difficult. They always ask me nicely so I always end up saying yes. What can I say? I'm a sucker for politeness.

6 p.m.

Done by the officials. Again! Jeezoz Christ. What a foking surprise. This one really does make no sense whatsoever. Kevin got shunted from behind by Bottas at the first chicane on the opening lap. Not only did it cause damage to his diffuser, costing him performance, but the stewards chose to give him a five-second penalty for leaving the track and then overtaking another driver. They didn't give Bottas a penalty. The one who foking shunted him. Apparently that was a racing incident whereas what Kevin did – or should I say what Kevin had done to him – wasn't. So the guy who has his race destroyed through no fault of his own gets a five-second penalty and the guy who runs into him and causes it gets off scot-free? We spoke to the stewards after the race and they said that the reason they gave him a penalty was because Kevin had gained an advantage. But he wasn't looking for an advantage! That's like finding a stolen car outside somebody's house and then charging them for stealing it even when you know they haven't. It's completely nonsensical! When we asked them why Bottas didn't get a penalty they said it was because it was a first-lap incident. What was ours, then? Lap foking ten? I even asked members of the team if I was missing something, just in case. The scary thing is that there are four stewards on the panel and so four people thought that this was a correct decision. Foking incredible! After that I think that Kevin got a little bit demotivated and he couldn't find a good balance.

Mick did really well today but he too had some bad luck. He managed a long first stint before coming in on lap thirty-three for soft tyres. He was down in eighteenth when he came out but within a few laps he was up to twelfth. Seriously, he drove the shit out of that car. His tyres were also in good shape, whereas the tyres on the cars around him were dying. He was on course for some points, for sure. Then, on lap forty-seven, Ricciardo had engine failure, which led to a yellow. Racing didn't resume and so that was it. Despite that, Mick drove really well this weekend and without hardly any laps beforehand. I'm pleased for him and he should be proud of himself.

The only other positive, I suppose, is that we're still a point ahead of AlphaTauri. A few weeks ago, I was talking about finishing seventh in the Constructors' Championship and now we're looking at eighth. If we don't manage that it would be a big disappointment. There are six races to go and we have to make them count. We need to have a clean race in Japan and we need to have a clean race in Brazil. They're the two races where we should be scoring points. Japan especially. There are lots of high-speed corners there and that's where our car is at its best. The other four races should be OK, but those two present the biggest opportunity. We cannot make any mistakes.

I have two weekends off now and I tell you what, I'm about foking ready for it. It isn't usually the case but the last six months have been hard going for me. Maybe I'm getting too old for it? And maybe that's a load of bullshit! A little bit of R&R and I'll be fighting fit again and ready for anything. Fights with the FIA, penalties, illegal invasions of sovereign countries, fights with the FIA, spoiled billionaires, the German media, fights with the FIA. You know the kind of thing. Or the kind of thing I have to deal with every foking day. I'll say it again, who'd have my life!

Tuesday, 27 September 2022 – Haas F1, Kannapolis, North Carolina, USA

2 p.m.

I'm sitting here in my office at Haas HQ in Kannapolis and, apart from one accounts person, I'm the only one here. Since the pandemic, everyone has started working from home. So why are we paying rent on this place, that's what I'd like to know? Let's scale down. The only problem we have there is that Gene owns the building privately and we pay rent to him. I don't think he'd be too happy if I suddenly served him notice. I'd be the one working from home then. Foking permanently! I think I might just park this problem and move on to something else.

When I arrived home after Monza I'd been away for the best part of three months and, although I'd intended to get some rest and recuperation, I've ended up foking firefighting and catching up with shit. You know, invoices and jobs that need doing. You forget how much of this stuff there is to do sometimes and when you're out of practice like I am it takes all day just to sort out a few small things. The part I find most difficult is switching from F1 mode to domestic mode. F1 mode is where I am for most of the time and no matter how hard I try it keeps on pulling me back. That's why everything at home takes me so foking long. It's infuriating. Why don't we just live in a caravan or something?

I shouldn't be complaining really, as over the last two or three weeks I've been able to spend some time with Greta, which has been great. I take her to German school at the weekends when I'm home and then we go and have lunch together. That alone is worth a week of sitting around and scratching my arse. What's that saying? It's the little things that matter. That's true for me.

As much as I enjoy spending time with Greta and Gertie, I always feel a sense of relief when I step back into F1. Not because I miss it, necessarily. Although I do sometimes. It's because when I'm away from F1 I leave certain things to chance so when I get a grip on it again I feel kind of relieved. I know a lot of people who experience something similar and they're not all control freaks. They're just passionate about what they do or about what floats their boat. Gene might own the team but I *live* the team and, together with Gertie and Greta, it's everything to me. It's true. They're the last things I think about when I go to bed and the first things I think about when I get up. Gertie, Greta and a team of two hundred people scattered all over the world.

So, what's been happening in the world of Haas lately? You mean apart from disappointment, penalties and a shit sandwich? Well, we're getting ever closer to sorting out the title sponsorship. In fact, I'm 99 per cent sure we'll have it sorted in time for Austin, which means we'll be able to celebrate and make the announcement there. You might think I'm weird for getting this excited about a sponsorship, but if this comes off it'll help to secure the future of Haas for many years to come. And they're the right fit, too, which I already mentioned. And they're Nasdaq listed. I found out the other day that they've been talking to other teams. They're perfectly entitled to do that and they've been very honest about it. That's the only thing that could mess it up now, I think. I'd be very disappointed if it happens.

The other thing to mention is silly season, which is ongoing. Gene and I still haven't made up our minds about Mick and we're not likely to do so until towards the end of the season. The rumour mill has been going into overdrive. According to the media, I've been talking to Daniel Ricciardo. What, about who's got the biggest foking nose? We know he'll be leaving McLaren at the end of the season so I understand the speculation. Some of the rumours, though! Apparently we've been sending each other WhatsApp messages. Jeezoz, thanks for letting me know, guys, because I didn't see any! I think I've talked to four or five drivers up to now. If Mick comes good nobody – apart from him – will be happier than me. And I want that to happen. Partly because I think he's a good guy and is talented but also because I cannot be bothered going through the shit of speaking to managers and hiring a new driver. There, I said it. They all get on my foking nerves, to be honest. If Mick starts scoring points on even a semi-regular basis and doesn't write off any more cars, I'll be happy to ask him to sign a new contract. And I might even give him a small bonus for saving me from having to speak to any more drivers and their managers.

OK, I'm catching a flight to LA soon for my meeting with Gene and the board. Then it'll be on to Singapore.

Saturday, 1 October 2022 – Marina Bay Street Circuit, Singapore

6 p.m.

Bearing in mind how things have been going for us lately, if I said to you that we've had quite a clean weekend so far you'd probably think I'd been taking drugs. Well, you can test me if you like because I haven't, and we have. Had a clean weekend so far, I mean. We haven't been taking drugs. Free Practice went well yesterday. Mick hasn't driven here before and Kevin hasn't driven here since 2019 so it was important we got the laps in. And we did. In fact, I'd say that both sessions were pretty much flawless. That's not a word that gets used very much in F1. I could get used to it, though.

And then on to today.

Free Practice was a washout, which meant the track was still changeable for Q1. Both drivers managed to benefit from this so it was hello Q2. Kevin breezed through in P7 and Mick squeezed through in P14. Unfortunately Mick didn't make it through Q2 but, given that he's never driven here before, he's had a good debut so far. Kevin did what Mick did in Q1 and just managed to squeeze through. He finished the session P10 and made up one place in Q3, which puts him P9 on the grid tomorrow. I don't want to jinx what we've achieved so far this weekend but given the position we are in we should be scoring some points tomorrow. We haven't managed to do that for the last five races and to go six without scoring any would be bad news. OK, that's all I can tell you really. As I said, it's been a flawless weekend so far. Let's hope it continues.

Sunday, 2 October 2022 – Marina Bay Street Circuit, Singapore

6 p.m.

Me and my big foking mouth! I say I hope I don't jinx things by declaring it's been flawless this weekend so far, and what happens? The whole thing turns to shit. Kevin ran into Verstappen, which basically finished his race. That's three times now. Twice with Hamilton and once with Verstappen. Dude, why can't you pick a fight with a car that isn't faster than ours? I remember saying to Ayao, 'The only thing that could make it worse now is if they issued a black and orange flag.' 'They won't do that,' he said. 'The damage to the front wing is minimal. Nothing's hanging off and there are no safety concerns.' 'What difference does that make?' I said. 'These guys don't know how to practise common sense most of the time.' I was kind of being facetious, as I too didn't believe they would issue one. Five minutes later, though, we get told that they *have* issued a black and orange flag. Jeezoz Christ! The part that was damaged was the endplate, for God's sake. And it's tethered to the foking car! It just makes no sense. He finished the race in twelfth, which wasn't too bad, but imagine what might have happened had he not made contact with Verstappen.

Mick's luck didn't go much better, unfortunately. He too made contact with another driver – Russell – which led to a puncture. He was running tenth at the time and was in a good position for points. After coming in, he lost some places and ended finishing one place behind Kevin in thirteenth. So, that's now six races on the bounce without any points. I'm seriously pissed off right now, and not just with the race officials. I'm going to go before I say something I regret.

Thursday, 6 October 2022 – Team Hotel, Suzuka City, Japan

12 p.m.

I landed about three hours ago and as I got into the back of the car taking me to the hotel I had a big flashback to 2002. I'd just arrived in Osaka with Niki for the Japanese Grand Prix. It was the last race of a pretty appalling season and we were both a little bit fractious. Actually, why don't I say it as it is? We were both as grumpy as fok! It must have been about eleven o'clock at night when we landed and it was raining like you wouldn't believe. Shit, we were miserable.

The guy who was taking us to the hotel was very formal and respectful. He was wearing gloves and a cap and when he saw Niki and me he bowed and then insisted on taking our luggage. 'I'm OK,' said Niki, putting both hands on his case. The guy didn't speak English so still went to take it and for a few seconds a tussle took place. There was only going to be one winner, though. 'I said I'm OK,' said Niki, grabbing his case as if his life depended on it. 'Just take us to the car.'

I was more than happy to let the driver take my case. It had been a shit flight and I was exhausted. Given what the rest of the season had been like (the only high point had been the freak podium in Monza, which we got two years on the bounce), we felt like a couple of convicts making our way to the foking electric chair. To be fair, we probably also looked like a couple of convicts making their way to the electric chair!

Niki Lauda hated being driven by other people and if you wanted to avoid an incident it was always best to just get him to where he wanted to be as quickly as possible. Don't fok about, in other words. Unfortunately, the man driving us had an Olympic medal in foking about. As well as staying in the slow lane no matter what, he kept on touching the brakes about every five or ten seconds and was on and off the throttle all the time. It was actually really uncomfortable.

If that wasn't bad enough, we were being passed by every single truck on the motorway and as they passed us a few hundred gallons of rainwater would get sprayed over the car. Niki had chosen to sit on the right-hand side of the back seat, behind the driver, which meant that when it happened he bore the brunt. After a while, he started marking each drenching by shutting his eyes and clenching his fists. I remember thinking to myself, *This man is going to foking explode soon!*

I realized after a while that the reason the guy drove like this wasn't because he was careful and cautious. He was just a really shit driver and had zero foking confidence. Knowing this was very comforting! I didn't say anything to Niki, though, as it might have pushed him over the edge.

After about twenty minutes, Niki started voicing his disapproval of this guy and he had reached exactly the same conclusion as me. 'This idiot cannot foking drive, Guenther. It is ridiculous. We'll never get there at this rate.'

Niki went on like this for the next few minutes until we eventually came off the motorway and arrived at a toll station. Seeing his opportunity Niki jumped out of the car, opened the driver's door and ordered the man to get out. 'I will drive us the rest of the way. You are incompetent. Come on, get out!' He then gestured for the man to shift his ass from the driver's seat but the man was going absolutely nowhere. 'Come on,' repeated Niki. 'Get the fuck out. I will drive us to the hotel. You are not capable of doing this.'

In an act of defiance, the driver grabbed hold of the steering wheel with both hands and then leant forward and started foking hugging it! I thought, *What the hell is happening here?*

Niki was not impressed.

'Come on,' he said, tapping the man on the shoulder. 'Do not be stupid. Get off the steering wheel and go and sit in the back with Guenther. You are not driving any more today.'

In yet another act of defiance the driver then closed his eyes, turned his head and faced away from Niki. He may as well have insulted Niki's mother because he was not happy. Shit, was he pissed off!

While all this was happening, a queue had formed behind us. Niki was obviously pretty recognizable and so, while he was berating the driver and trying to move him to the back seat, the people in the queue started getting out of their cars to watch. Thank fok it wasn't these days because everyone would have been filming it. Including the driver, probably! The standoff lasted for about five minutes until I eventually decided to get off my ass and try and make matters worse. It had lost its appeal now and I wanted to move.

'Come on, Niki,' I said. 'Look at all the people watching. Let's just get in the back and let him drive.'

'Let him drive? But he cannot fucking drive. He has already dem-
onstrated this.'

With that, Niki made one last attempt to get the driver to change
seats. 'Come on, you shithead,' he said, trying to prise the driver's
hands from the wheel. 'Get in the back, you clown!'

I said, 'It's not going to work, Niki. The driver would obviously
rather die than let go of the foking wheel.'

Niki might have been beaten but he was going to have the last
word.

'Look at him,' he said, pointing at the driver. 'He is holding on
to the steering wheel like a child would hold on to a fucking teddy
bear. This is not normal behaviour.'

'Neither is standing there and telling him to get in the back!
Come on, Niki. We look like a couple of wankers.'

'I am not a wanker, Guenther. *He* is a fucking wanker!'

Friday, 7 October 2022 – Suzuka International Racing Course, Suzuka, Japan

6 p.m.

What a start to the weekend. Mick crashed his car earlier during
FP1, causing about $700,000 worth of damage. Yep. Wing gone.
Nose gone. Underfloor gone. And the chassis! You should see the
car. It's a foking mess. Worse still, it happened on the foking in-lap.
On the in-lap! Sure, it was very wet out there on the track, but
nobody else managed to write off a car while they were driving
carefully back to the pits. What did I say before? We have to have a

clean race in Japan. And what happens? We lose a car after five minutes and now have to build another. As well as affecting the budget,
this will also affect the confidence of the team as we'll be starting
the weekend from a reduced position. Mick has never raced here
before and he will now miss a complete session. I cannot have a
driver who I am not confident can take a car around safely on a slow
lap. It's just foking ridiculous. I'm trying not to be too hard on the
guy but right now I am seriously pissed off. How many people
could we employ with $700,000? And I have to now find that
money from somewhere. It doesn't just materialize. I haven't told
Gene yet. That's going to be an interesting conversation. I know
what he'll say. 'Why did you send him out if it was raining?' I'm not
looking forward to that one.

With Kevin, things went a little better, in that he managed to get
through both sessions in one piece. He also did some testing, which
gave us some interesting and useful data. I'm trying to find some
positives here but there aren't many around at the moment. In fact,
and this is going to sound controversial I think, I would say that at
the moment the only weak link we have at Haas F1 is the drivers.
The rest of the team have been performing very well and they are
not making mistakes. I'm not saying that they are perfect (and I am
certainly not), but they have all done their jobs well and to the best
of their ability. But it is not just crashing cars and crashing into
other drivers that is the problem here. The form of the drivers in
general has been suffering lately and there is no excuse for that.
After two years of shit we finally have a good car. It's frustrating.

People are asking me now if I need to have strong words with
Kevin and Mick about this. What, like I did with Kevin and Romain
that time? All that did was give Netflix some more viewers and give

me a foking catchphrase. 'He does not fok smash my door!' I'll take that to my foking grave. I could even have it as my epitaph. 'I came, I saw, he fok smashed my door.' Jeezoz.

In this situation, shouting at them isn't going to solve anything, for sure. But neither is even speaking to them. I've tried the arm around the shoulder thing and, if anything, it just makes matters worse. It makes them think that you feel sorry for them and it can almost create a kind of victim mentality. In order to improve they have to come to the conclusion themselves that they – not the team – is what is separating us from the points. It might sound like I'm being hard on them but the results don't lie and we are not where we should be. In Monza I said that in order to be a top driver you need promise, talent, ability and experience. In hindsight I think that I missed two out because what you also need is dedication and the right attitude. That, in my opinion, is perhaps what is lacking a bit with our drivers. I don't actually think it's intentional. It's just what separates the good drivers from the very top drivers. Look at Alonso. He is older than God, yet every week he gives 100 per cent and is always motivated. Even if a situation is shit he will squeeze something good out of it and will always push his team to improve. He's a talisman. A sometimes grumpy one, but a talisman all the same. And he doesn't have to do this, you know. He must have at least a thousand euros stashed away in the bank and he could just go and live on an island somewhere. That isn't what he wants, though. He obviously feels he isn't done yet and, by the look of his performances this year, he's foking right! For me, he is the benchmark when it comes to what you need to be a complete driver. He's relentless.

Look, I'm sure this is only temporary with Mick and Kevin. It's starting to get expensive, though, and at the moment I'm feeling permanently apprehensive. If we get nothing out of this weekend, that will make it seven races in a row without us scoring points. That's almost a third of the entire season. If that happens, I think we can say goodbye to seventh in the Constructors'. Especially if Aston Martin score points.

Anyway, happy foking Friday, everyone!

Saturday, 8 October 2022 – Suzuka International Racing Course, Suzuka, Japan

6 p.m.

One word for today's performance. Lacklustre. Kevin went out in Q1, which was very disappointing, and, although Mick made it through to Q2, he finished last. They qualified eighteenth and fifteenth respectively, which is shit. Had this been earlier in the season I might have been using more positive language but at the moment I can't. Japan is one of our biggest opportunities and so far we are doing a very good job of foking it up. I don't know what else to say really. This is probably my first low point of the season. I can't see the wood for the trees.

I had a conversation with Gene earlier about what happened on the track yesterday. He was very quiet. Even quieter than usual. What is there to say, though? At least he didn't have a go at me for sending Mick out in the wet. He agrees that an F1 driver should be able to bring a car in on a slow lap safely, regardless of the conditions. Even I was a bit lost for words. Usually I'm a master at filling in the gaps and avoiding long silences. Not today. It wasn't good. We need to get things back on track. Preferably by tomorrow!

Sunday, 9 October 2022 – Suzuka International Racing Course, Suzuka, Japan

8 p.m.

Well, that was truly shit. At least it was not the drivers' fault this time. In fact, if anyone is to blame for what happened today, it's me. I'm the one responsible for the team and today the team did not perform. Our decision making was slow and far too careful. In fact, we weren't even smart enough to do what the other teams did. We discussed it, but we didn't do it. And what is the worst decision you can make in life? No decision. Had we come in when Vettel and Latifi did we would have been flying. We weren't brave enough and, by the time we realized, it was too late. Maybe this too is a hangover from last season? I don't know. What I do know is that we let ourselves down.

After the race I wasn't even angry. I was just disappointed. Disappointed in them but especially disappointed in me. I thought about calling a team meeting but it would have been too foking depressing. And I've decided that enough is enough. Changes will be made because of this and I've already sent a message to Ayao to put the wheels in motion. I'm afraid I cannot elaborate on it but something like this will not happen again at Haas. No way.

The only bright spot at the moment is that the next race is Austin. Our home Grand Prix. In normal circumstances, I would be really looking forward to it but at the moment I am a bit flat. I need to pull myself together as, with any luck, we will be unveiling our new title sponsor there. Actually, even just writing that makes me feel a little bit better. It reminds me that we are not defined by the result this weekend and we actually have a future. Good times are ahead of us. I'm sure of it. Foking hell, I must look and sound crazy at the moment. One minute I'm pissed off, and the next I'm ultra-positive. Would the real Guenther Steiner please stand the fok up!

Sometimes in life you have to be your own best friend and over the years it's something I've learned to be quite good at. If nobody else is going to make you smell the coffee, why not do it yourself? You know, I'm starting to feel a bit more like my old self again. We still have four races left until the end of the season, which means four chances of scoring some points and giving ourselves something to celebrate. One of the things I love most about Formula 1 is that you never know what is going to happen next. Sure, it can kick you in the ass sometimes. Don't I foking know it! But it is also part of what keeps us dreaming.

Anyway, right now all I want to do is go home and see my family. I just wish they weren't 7,000 miles away!

Monday, 17 October 2022 – Steiner Ranch, North Carolina, USA

9 a.m.

There is something I forgot to tell you about from Japan. It happened on race day. I was sitting on the pit wall watching the coverage before the race when, all of a sudden, the cameras went into the crowd and found a woman holding up a sign saying, 'Guenther, will you marry me?' A cameraman was standing right next to me at the time so caught my reaction. And I bet David Croft and Martin Brundle were taking the piss! She was sitting next to a guy wearing a mask with my face on, so what was she wanting? Two Guenthers? That's just sick. Where the hell did she escape from?

Let me dispel a little myth for you. People think that life on the pit wall is all data, strategy and concentration. That's not concentration. It's boredom! Ayao and Pete Crolla have the attention spans of a foking three-year-old, so after a few laps they'll start talking about what they're having for dinner or which bar they are going to. It's the same all over the pit lane. Fred talks about France all the foking time and Toto talks about how much money he's made since he sat down. And don't be taken in by all the stern and concerned faces. If somebody on a pit wall looks like that it's because they need the toilet and are trying to hold it in.

Since arriving back from Japan, almost every waking moment has been dedicated to getting the new title sponsor over the line. The money involved is massive (almost a nine-figure sum in total) so, although it's taking a bit longer than I expected, we just have to sit down and be patient. The good thing is that, because it is an American company who are actually based in Texas, we're all keen to announce the partnership in Austin and have called a press conference for Thursday. That gives us just three days to sign the contracts. It should be OK, though. We're at the fine-tuning stage now so it's as good as done. Gene and the board are very happy, and if they're happy, so am I.

2 p.m.

Ever since we announced the press conference for Thursday, the press and media guys have been going foking crazy about what we will be saying. They think it has something to do with the drivers and at the moment I'm happy for them to believe that. I should put a rumour out there that we're considering replacing both drivers. That would give them something to think about!

All of the big websites have fallen for it. Even PlanetF1. They said, 'The press conference is set to take place on 20 October and PlanetF1 report Haas have not yet revealed who will be at the press conference, suggesting the team could be ready to reveal whether or not Schumacher will still be partnering Kevin Magnussen next season.'

Not this time, suckers! If you google Haas F1 at the moment, almost every story is about the driver situation. There are literally hundreds of stories. And that's precisely what they are – stories! It's all speculation. You know what, it's nice to be in control for a change. If they actually knew the truth about the driver situation at Haas they would be camping outside my foking office, as I've already spoken to enough drivers to fill a bus.

I'm hoping that because of all the shit with Uralkali at the start of the season, the press will still be interested when they find out what we want to tell them. It's the biggest sponsorship deal in F1 this year so that should count for something. I cannot tell the press this or go into detail here but another team has even tried poaching this sponsor from us by undercutting the investment by a third. Cheeky bastards! The sponsor wasn't interested, though. They know we're the right team for them.

OK, I fly to Austin in a few hours so I'd better get my shit together. It's going to be an interesting week.

Wednesday, 19 October 2022 – Circuit of the Americas, Austin, Texas, USA

The contracts have just been signed so we're all good to go. Our new title sponsor is MoneyGram, I can announce. They're pretty well known globally but are an American institution. They're also not very corporate and the guy who's in charge is great. We're going to have fun with them and hopefully some success.

More good news! Mick's uncle has been coming out with more shit again. According to him, everybody at Haas is doing a good job except for me and Gene. That's right. Despite the fact that his nephew has written off two cars already this season and hasn't scored any points yet, Guenther Steiner and Gene Haas are the weakest links. The man is obviously a genius.

Surely the sensible thing for him to do in this situation would be to see if he could help his nephew and bring something to the table, instead of just trying to stir shit all the time and get headlines.

Thursday, 20 October 2022 – Circuit of the Americas, Austin, Texas, USA

When the press and media guys asked Stuart what the press conference was about this morning we thought some of them might not come. After all, they're not obliged to cover it. They can write about what they like. Then, when I turned up to the venue with Stuart and Greg and Alex from MoneyGram, I couldn't believe it. The place was full! The guys were also interested in what we had to say and seemed genuinely pleased that we, the smallest team on the grid, had found a new and credible title sponsor.

I think already this has rehabilitated our reputation post Uralkali. The invasion might not have been our fault, but shit associations like that stick and it kept us in the headlines for all the wrong reasons. Now we have a title sponsor that is American through and through, Nasdaq listed, and is serious about helping us to be successful. If I were a marketing person I would probably have said, 'serious about joining Haas on their journey to success', but fok it. No more Russians, though.

Friday, 21 October 2022 – Team Hotel, Austin, Texas, USA

7 a.m.

I received some tragic news last night. Harvey Cook, a mechanic who had been with Haas a long time, passed away yesterday after a long battle with cancer. Harvey joined us about five years ago and was one of these kids who would do anything for the team. There was never a 'No' and never a grumpy day with him. He just loved it and was Haas through and through.

I used to call him the goat because he could jump over tables and tensator barriers from a standing start. And he could climb walls. Seriously, like Spiderman. He can't have weighed more than about

50 kilos and was just skin, bones and muscle. He was a complete freak of nature and would have us all scratching our heads.

I remember he had a few health issues in his early days with Haas but then about two years ago I was told that he had cancer and had only six months to live. It was tragic news and everyone was devastated. Apart from supporting Harvey financially, we thought, *What can we do for him and his family?* His wife, who was then his girlfriend, was a big Formula 1 fan like Harvey, so we got them both to Silverstone and gave them a good day. I think Harvey had already beaten the doctor's prognosis by this point and was very upbeat. 'Well, I am still here,' he said. 'I'm still breathing.'

Amazingly, Harvey carried on defying the doctor's predictions and with the help of some new drugs, I think, he started getting better. 'Jeezoz Christ, you're supposed to be foking dead!' I said to him one day. 'I know!' he replied. 'But you don't get rid of me that easily.'

Then, during the off-season last year, he surprised us even more by asking if he could come back to work. 'Are you serious?' we said. We honestly couldn't believe it. But of course we said yes. One thing you never, ever do is kick a man while he's down. That is an important principle of life for me. Even if Harvey had been well enough only to make the tea at a race, he'd have been welcome. He did a lot more than that, though. He more than pulled his weight and was an asset as always.

Then, about three weeks ago, I was informed that he'd had a very bad seizure. Even at this point I didn't take it too seriously because by now Harvey had become almost invincible in my eyes. Death kept on challenging him and he kept on winning. Take that, you bastard! Harvey will always get better. A few days later he had another seizure, which left him unable to speak, and then yesterday he passed away. He was only thirty-one years old.

As a tribute to Harvey, we're going to carry his name on both cars this weekend and at some point I'll gather the team together and we'll give him a one-minute round of applause. Some people do a one-minute silence but Harvey wouldn't want that, I'm sure. He liked noise!

Nothing else to talk about today. This entry is only about Harvey.

Saturday, 22 October 2022 – Circuit of the Americas, Austin, Texas, USA

10 a.m.

I know we were all impressed by Miami this year but if today and yesterday are anything to go by, Austin could be even more impressive. The atmosphere is like a festival and I've been told that they're expecting anything up to 450,000 fans over the weekend. That's even bigger than Melbourne and Silverstone. I think they also have Green Day and Ed Sheeran playing. I'd prefer ABBA, but apparently they weren't available. There also seems to be more of a buzz about us being the only American team this year. More, please! That's one of our proudest USPs and, to be brutally honest, it's something we haven't either enjoyed or exploited enough over the years. The MoneyGram partnership will help that, I think, and we've really gone to town this year on the car livery (which looks amazing) and the photoshoots. The drivers were photographed wearing cowboy hats with their newly designed race suits. It was pretty cool and they had fun. Kevin would like to have had some cowboy boots too, I think, to make him a bit taller. This is our identity, though, and it should be celebrated as often as possible. Definitely more to come. Even I have dual citizenship now. We're American and proud!

1 p.m.

Fred from Alfa and Mario Isola from Pirelli ambushed me during the press conference earlier, wearing T-shirts with my foking face on! T-shirts, incidentally, that according to Stuart are available from the Haas website. Some people in the crowd were spotted wearing them yesterday, I think, and so Fred and Mario decided to follow the fashion. They also have slogans on with bad words that I allegedly said. They looked quite good, to be honest. I was impressed. I don't know whether this is true or not but I was told that when we started selling the T-shirts the website crashed for four hours. Is that a good thing or a bad thing? I don't know. Management style.

6 p.m.

The weekend has been very sketchy so far. Antonio Giovinazzi went out in FP1 in place of Kevin yesterday and crashed after four laps. He was hit by a freak gust of wind, and despite managing to get the car back to the garage, he'd burnt the clutch out in the process. This meant that Kevin couldn't test the new prototype compounds in FP2, which wasn't good. It wasn't a good day really, and then qualifying today didn't go much better. It looked like we could make it into Q2 at least, but it didn't work out. Both out in Q1. Kevin starts P15 and Mick P18. Nobody ever likes going out in Q1 but this time I'm not too despondent. In fact, we have a good chance of making it up the field and I wouldn't be surprised if we scored a few points.

What is this, you must be thinking. Positive Guenther? Surely he is an imposter. Well, there's nothing like a home Grand Prix, a fast car, a new title sponsor and a few hundred thousand fellow Americans to prevent you from being negative. I also want to do it for Harvey. We all do. And we can.

Sunday, 23 October 2022 – Circuit of the Americas, Austin, Texas, USA

10 a.m.

I was right about the crowds. According to Stefano, the attendance this weekend is about 440,000, which is the biggest of the year. I know we have a few races left but nobody will pass that. Nowhere near. The guys from MoneyGram have been very impressed so far. I should tell them that not every Grand Prix will be like this, though. One day, maybe.

5 p.m.

You remember what happened in Miami when Kevin wanted to change tyres and it didn't work out? Well, the opposite kind of happened today really. Kevin got in a good position to score points and about ten laps from the end we thought it would be a good idea for him to come in and change from mediums on to softs. 'I think the ones I'm on are OK, guys,' he said. I remember looking at Ayao. My instinct was just tell him to come in as I didn't think the tyres would last, and if they didn't there was a chance he wouldn't finish. 'Seriously, they're OK,' he said. 'Let me stay out.' The chances are he would have retained his position but you never know what's going to happen with a pit stop. The vast majority go without a hitch but this is Haas, remember. And remember that foking jack! Anyway, that turned out to be the right thing to do as the tyres were OK and in the end he finished ninth. You see. I told you I was right to be positive.

The final lap of the race was one of the best I've seen all season in terms of racing. Kevin and Vettel were each vying for eighth position and the battle between the two of them was amazing. Although Vettel managed to overtake Kevin on the final corner (*scheisse!*), he said afterwards that it was some of the best racing he's ever been involved in. I don't know Sebastian very well but he's been a great ambassador for the sport and obviously a great champion. Formula 1 will miss him, for sure. Apparently he does quite a good impression of me. I haven't heard it yet but one day I hope I will. Cheeky German bastard.

Poor Mick had an unlucky race in comparison and he's rightly disappointed. He went over some debris, which damaged his car, and he lost about forty points of downforce. You cannot do anything when that happens and so for the rest of the race he was just sliding around. Poor kid.

I'm being positive *and* sympathetic. What the fok is going on?

10.15 p.m.

Stop Press!

Jeezoz Christ. More has happened since the race than during it. About five minutes after the chequered flag, I was told that Alonso, who finished seventh, drove the final few laps of the race without his right mirror. You cannot do that legally and it should have been spotted by the stewards. I've lost count of the number of times we have received black and orange flags for such things. What can I say? Certain drivers seem to be 'luckier' than us. Seriously, though, Haas are foking famous for getting penalized so when it came to light that the stewards had 'missed' what happened with Alonso, we decided to put in an official protest. Sure, I got a load of shit from Otmar when he found out, who is the team principal at Alpine, but I don't care about that. Remember those little battles I told you about that happen within F1? Here is your example.

Under normal circumstances, our team manager, Pete Crolla, would put in a protest like this on behalf of Haas but he wasn't available. This meant that I had to do it instead and let's just say that objecting to something calmly and clearly isn't really one of my good points! I've done it before in front of the stewards and every time I've been fined for losing my temper and swearing.

In the end I took Ayao with me, who is very calm and Zen. He's like a walking Valium tablet really, and every time I got excited I could feel his calming influence beside me. Anyway, the verdict is due any moment now.

10.35 p.m.

We just received the written verdict, which says that the stewards were deeply concerned that Car 14 [Alonso] was not given the black and orange flag, or at least a radio call to rectify the situation, despite two calls to Race Control. 'Notwithstanding the above,' it says, 'Article 3.2 of the Formula 1 Sporting Regulations is clear – a car must be in a safe condition throughout a race, and in this case, Car 14 was not. This is a responsibility of the Alpine Team.'

Alonso has now been dropped from P7 to P15, which moves Kevin up to P8. Despite the extra two points, this is more about consistency for me. Three times we have been made the foking whipping boys with black and orange flags and it's not fair. I'm not trying to gain an advantage here. As long as I'm given fairness, I can fight on my own. I don't need any help.

Thursday, 27 October 2022 – Team Hotel, Mexico City, Mexico

8 a.m.

Yesterday I had a very long meeting with Gene about the driver situation. To be honest, we're still undecided. Although it has to be said that Mick didn't do himself any favours in Austin. That's eight races without a point now. I can't see that changing here. That is not just pessimism talking. That is realism based on what I've just said and the fact that, for some reason, we've always been shit at this circuit. Since 2016 we have only scored points once here and with one driver. That was Kevin back in 2017.

One thing I've noticed is that everybody is quite tired now. As the season goes on your skin gets naturally thinner and your stress levels start to increase. Although this year, because of everything that has happened in our world, it has been on a slightly different level for us. Even I am really feeling it now and usually I'm the last one to drop. I think the growth of Formula 1 has also been a contributing factor. It isn't just Haas who are tired. Everybody is. The teams, the media. Everybody who travels with this circus of ours is exhausted. Anyway, I'd better get my shit together and get to the circuit. I've got about ten interviews to do and a couple of Q&As.

Oh, I almost forgot. Guess what? Tomorrow there will be a meeting involving the teams, the FIA and F1 about – you guessed it – black and orange foking flags. The situation has come to a head now and unless I get told afterwards (the team managers are going and not the team principals) that something has changed and that the FIA and us are going to have a serious discussion about the damage that has already been done, I will not be happy. I'm expecting it to be constructive, though. Everybody knows that something has to change, the same as everybody knows that Haas have been screwed over unfairly three times already. Watch this space.

Friday, 28 October 2022 – Autódromo Hermanos Rodríguez, Mexico City, Mexico

9.30 a.m.

I really don't know how to put this without sounding like a big head, but ever since arriving in Mexico I haven't been able to go anywhere without being mobbed by people. I am even getting chased now when I want to go to the toilet! That has never happened before. The growth of Formula 1 has been pretty brutal and I suppose things like this are the result. It's getting to a point where it's not really that nice sometimes and can also be a bit scary. And it's also tiring, to be honest. Being surrounded by hundreds of people several times a day who are all vying for your attention and want a selfie takes it out of you. Not just physically, but mentally, too. Even in the paddock now it is sometimes no different. As I said, going to the toilet if you're a driver, or even a Guenther, can be a dangerous business and it has to be planned almost militarily. You wait for a quiet time and then run for your foking life and pray that you don't have to queue!

Depending on which circuit you are at, the toilets can sometimes be anything up to a hundred yards away and so in some cases you don't stand a chance. And that's the thing I don't like about it. Nothing is natural any more. You have to think about every move you make these days. I'm sure I will get used to it, but at the moment it feels weird. The good thing is that almost everybody who approaches me, regardless of the country or situation, is polite. They're a little bit too enthusiastic sometimes, but that's OK. You have to rise above these things. We certainly cannot start asking people to go easy on us just because we're twenty races into a season and are feeling tired.

Can you imagine? 'Please, Mexico. Go easy on poor Guenther. He's a little bit fatigued.' Every Grand Prix is special to the people of the country where it takes place. Tired or not, we have to give them all exactly the same treatment and experience. And we will.

Oh, by the way. I spoke to Stefano earlier and apparently they're expecting about four hundred thousand people this weekend. Four hundred thousand crazy Mexicans! Somebody, pass me the tequila!

12 p.m.

I was talking to Pete Crolla about our history in Mexico earlier and he reminded me that one of the reasons we have always been shit here is to do with the altitude. It's high, so sometimes during a race we have had to open our cooling system. And, when you open up the cooling system on an F1 car, you lose downforce. I'm trying to put it in layman's terms here so it makes sense to everybody. This year the cars have larger cooling systems so it should be OK. Who knows, we might even not be shit now!

6 p.m.

Free Practice went OK, I suppose. Pietro went out instead of Kevin in FP1 but he had a turbo problem so after three laps he had to come in. Mick had a good day, though, and managed to run through his entire test programme. He should be ready for tomorrow. Kevin managed to get most of his test programme done in FP2, despite an engine change – although there wasn't enough time for him to get out with the race tyres on. We can live with that.

Despite the cooling system issue being resolved, we're still not really where we should be here and we can't work out why. We're pretty sure it's still to do with altitude but we need to work on it and get it fixed. Which reminds me. I need to speak to the team in Italy about that. They are working on next year's car and the language coming out of there at the moment is pretty positive. No panic. No paranoia. What the fok is going on? It is less than four months to testing. How crazy is that?

Anyway, through my office door I can see about two hundred people hanging around outside our hospitality unit looking this way and holding a foking phone each. And I need to go to the toilet. Fok!

10 p.m.

OK, now I'm back at the team hotel I can tell you about the meeting that took place earlier about black and orange flags. The FIA explained that the reason mistakes had been made was down to two things: the difficulty in evaluating the level of damage, and the difficulty in keeping track of the state of twenty cars consistently. This, they said, had caused an uneven distribution of flags (you're foking telling me!), which had led to cars that were safe being deemed unsafe.

The upshot of the meeting was that, with immediate effect, the FIA would be limiting the use of black and orange flags to cases when a car has suffered significant damage to a structural component. They then gave several examples of what would not be considered suitable for a black and orange flag, which are: a broken front wing endplate, a faulty minor bodywork component such as a brake duct winglet or (this one wasn't much of a surprise) a damaged mirror, or a mirror that has flown off.

They finished off the meeting by saying that they would engage with the teams to discuss how the criteria could be refined for next season and hoped that in the meantime they had done enough to address it in the short term.

Yes and no.

The fact that it shouldn't happen again is obviously encouraging, but from our own point of view we now have to build a case about the damage that has already been done. I'm not going to let this lie. One fok up is bad enough but with three you are asking for troble. To set the wheels in motion, Pete has asked our head of strategy, Faïssal, to look at the respective race strategies and provide estimated finishing positions had the black and orange flags not been issued. It obviously won't prove anything, but it will be part of our case. I'll come back to you with the results.

Saturday, 29 October 2022 – Autódromo Hermanos Rodríguez, Mexico City, Mexico

10 a.m.

I believe that if we hadn't won those points in Austin, the team would be finding it really difficult to cope. I was a tiny bit worried when we arrived in Austin that they might not be able to make it through to the end of the season. I don't mean I was afraid that they would all start collapsing. When you're tired and are thinking of home, mistakes can start to happen. And especially when you've had a season like we are having. I know I'm probably repeating myself but it's true. We needed something to get us over the line and fortunately it happened.

Anyway, it took me almost half an hour to get through the entrance into the paddock earlier. Thank God I'm quite tall because if I was a shortarse like most of the drivers I'd have been trampled to foking death. How the hell does Yuki Tsunoda cope?

6 p.m.

The following two sentences, which sum up Free Practice, also sum up our season so far, I think. Kevin made it through to Q2, which is a first for Haas in Mexico. Then he got a penalty and starts P19. It was the same for Mick. He was on course for Q2 at the very least (Jeezoz, he was quick) but cut a corner so his time was disqualified. Swings and foking roundabouts, as usual! To say I'm disappointed would be an understatement but I'm too tired to complain any more. I'm going to walk to my car now, which will take me about four hours, and then I am going to drive back to the hotel, fall into bed and go to sleep. *Buenas noches.*

Sunday, 30 October 2022 – Autódromo Hermanos Rodríguez, Mexico City, Mexico

6 p.m.

No pace. Altitude? Who knows? Mick sixteenth. Kevin seventeenth. Bullshit!

Adiós.

Friday, 4 November 2022 – Charlotte Airport, North Carolina, USA

Tomorrow is the official launch event for the Las Vegas Grand Prix. They're expecting about forty thousand people to turn up. Forty thousand! Not bad for a minor sport. I don't know the full extent of what they've got planned but last week Stefano asked me if I would make an appearance, and I agreed. Gertie is coming with me. Not Greta, though. You honestly think I'd take my daughter to Vegas? I'm also going to be filming a short promotional film there for the Grand Prix, alongside Kevin. Well, not actually *alongside* Kevin. We're filming our parts separately and, as far as I know, he's already done his.

All I know about the event at the moment is that Lewis Hamilton, George Russell and Sergio Perez will drive their cars up and down the strip for a bit and then will be interviewed on the main stage by Naomi Schiff and David Croft. This is all gearing up for the main event, of course, which is me. After all that there will be a concert by a band called The Killers. I have no foking idea who they are but apparently they're quite good. Anyway, it'll be a bit of fun and Gertie is looking forward to it.

Saturday, 5 November 2022 – Caesars Palace, Las Vegas, Nevada, USA

2 p.m.

I've just finished doing some interviews in a suite here at Caesars Palace. It's pretty impressive. I thought I recognized it from the film *The Hangover* but apparently they recreated the suite in Hollywood somewhere. Bearing in mind what happened in the film, I'm not foking surprised! I also filmed my part for the promotional film here. They weren't lying about the numbers. Already there are people everywhere. It's foking crazy. Later on we have the cars and then the interviews on the main stage. I need a rest first. Followed by a foking big drink. I think I've earned it.

7 p.m.

Foking hell! I don't really know what I was expecting exactly but that was truly off the scale. The only problem I can see the drivers having at the Grand Prix next year is focusing on the race. There are going to be so many distractions for them. Not least a billion lights and several hundred thousand people getting completely shitfaced!

   The event was just incredible. They had drones everywhere letting off fireworks and the cars on the strip went down like a storm. I'm not sure how I'm going to cope next year with all the selfies. I got mobbed leaving the hotel and it took me almost ninety minutes to get done. It was foking crazy! Nice, but crazy. Then, when I was introduced on to the stage, everybody went mental. It was like Australia all over again. I'm an Italian wanker with a big nose, a weird accent and a German name. What the hell do these people see in me?

A couple of years ago somebody suggested that I should have a bodyguard at races and at first I was dead against it. 'What do I need a foking bodyguard for?' I said. 'I'm not Lewis Hamilton!' Then they said that in some parts of the world there is a danger of things like kidnappings. I already knew that but I never thought for a moment that I would be included. Can you imagine a gang of kidnappers going back to their boss after a hard day on the road?

'What did you get today, guys?'

'We got a Guenther.'

'What is it worth?'

'Fok all, by the looks of things!'

Last year in Brazil I had a bodyguard. He wasn't the sharpest tool in the box, though, and ended up dropping his gun in my car. I said, 'What the fok are you doing, dude? Are you trying to kill us both?' That put me off and the only time I have a bodyguard now is in Mexico, and only because I'm told to.

OK, Gertie and I are going out to dinner now at Gordon Ramsay's restaurant. Hell's Kitchen, I think it's called. I've met Gordon a few times now and he's a nice guy. Swears a bit too foking much, but nobody's perfect. I'll tell you one thing, I wouldn't mind being a dollar behind the Scottish blond-haired bastard!

Thursday, 10 November 2022 – Interlagos, São Paulo, Brazil

11 a.m.

I landed at São Paulo at 6.30 a.m. this morning after a pretty shitty night flight. It then took me over an hour to get through immigration and two hours to get from the airport to the track. Am I moaning? Too foking right I am! I'm exhausted. Mind you, what do you expect in a city with a population of about a billion people? It's crazy here. I love it, though. Brazilians have a real affiliation and understanding for Formula 1 and they're always very warm and welcoming. It's going to be a good weekend. Well, hopefully. I've said that before a few times.

Friday, 11 November 2022 – Interlagos, São Paulo, Brazil

7.50 a.m.

I always drive myself to the track in Brazil, which means I can come and go as I please. The only problem is the traffic. Although it's only about 8 miles from the team hotel to Interlagos, it can take well over an hour sometimes, especially on a weekday. Today being Friday, I'm expecting the worst. You never know. Maybe I'll be lucky just this once?

9.45 a.m.

Jeezoz Christ! That was the worst journey yet. One hour and three quarters! Just to rub salt into my foking wound, when I was stuck in traffic some street vendors kept trying to sell me Mercedes caps. Really? I said, 'I'm good, thank you very much, guys! And by the way, where are all the counterfeit Haas caps?'

It appears that *Drive to Survive* has become a bit more popular here as, when I arrived at the entrance, some people surrounded the car and started chanting my name. At first I thought they might be an angry mob but then I reminded myself that I don't work for Red Bull any more. Only joking!

When you actually get to Interlagos there are some stairs that take you from the car park up to the paddock and every year it takes me longer to get up them. I need one of those stairlift things. Anyway, fortunately there weren't too many selfie hunters at the entrance so I was in my office within a few minutes. What am I expecting today? I have no idea, to be honest. If we go by our history here, not very much. Double points in 2018 and apart from that, zilch. We should be quite fast but with the weather being so unpredictable, who knows? We're in a battle for eighth position in the Constructors' Championship at the moment with AlphaTauri so our main aim is to do better than they do. Just to put this into perspective, the difference in prize money between eighth and ninth in the Constructors' Championship is ten million bucks.

10 p.m.

WE GOT FOKING POLE!!!

When I said earlier that anything could happen, what I meant was that we could end up going out in Q1 or we could make it through to Q3. Well, we went one better than that today. We went right to the top, baby. It still feels pretty weird even saying it, let alone writing it. I couldn't wait to call Gene and he was over the foking moon. Quietly so, but he was very pleased. After that I called Gertie. She's obviously been with me every step of the way since we started Haas and by association she's been through every emotion I have. To a lesser degree, perhaps, but whatever I feel, she feels too.

Until today our best ever qualifying was at Imola this year, when Kevin got fourth. That was pretty amazing but this is on a different foking scale. You see things through new eyes when you get pole position. You exist in a different world and when you are there it opens up new dreams.

This isn't the first big success I have had while in charge of a Formula 1 team. You remember Jaguar that I ran with Niki Lauda? Well, in 2003 our driver Eddie Irvine finished third at Monza. The reason you might not be aware of this is because the car was the biggest piece of shit ever made. It was made before Niki and I arrived, though, and so basically we just went racing. I remember being with Niki and Eddie after the race. We all said, 'How the fok did that happen?'

The only thing the car had going for it — and I mean, the *only* thing — was that it had almost no downforce so on high-speed tracks like Monza it was OK. If it didn't break down, that is. At the time our only aim was to get through a race, so to end up on the podium? In the other sixteen races that season, Eddie had six finishes and ten retirements! That was always the problem with Jaguar, apart from all the other shit. In five seasons they had sixty-nine retirements. To put that into perspective, in seven seasons at Haas we have had fifty-five so far and with a lot more races.

I remember the presentation of that car like it was yesterday. In those days, launching an F1 car was always very spectacular, regardless of how shit it was. We launched ours at Jaguar's research and development base in Whiteley. We were also the first team to do it, so the press and media were all over the place. Just before the launch, Niki had a meeting with the technical people and when he came out he was absolutely foking fuming. 'You know that last year's car was shit, yes?' he said. 'Well, apparently this one will be even worse.' I have never seen a man as angry before in my entire life. Niki wanted blood! It wasn't just the fact that the team had made a step backwards that made him angry. Although that was about 95 per cent of it. He was about to go out in front of a hundred press people and tell them all what a foking great car it was, when all the time it is a shit heap! I remember Niki saying to me, 'Hey, Guenther. What the hell am I going to tell them?' I said, 'Tell them it's shit but that it will get better.' 'I don't want to *tell* them that the car is shit,' he said. 'That will spoil the foking surprise! No, I have decided. Let them find out for themselves.'

I keep forgetting that by the time this book is published everyone will know all about this. What I will try and do, then, is go through what happened from my own perspective and then try and explain what something like this means to a team like Haas.

OK, first up is qualifying. It had rained in between Free Practice and Q1 so both drivers went out on intermediates. I think Gasly was the first to change to softs and pretty soon everyone followed suit. Kevin's first lap had been deleted due to track limits but when he went out on softs he managed a 1:13.954, which took him through to Q2. Unfortunately Mick was almost three seconds behind Kevin and ended up P20. He thought the track was wetter than it was and underestimated the grip, I think. He was obviously frustrated, but so was I.

The weather conditions for Q2 remained mixed but, after coming in part way through the session for a fresh pair of softs, Kevin posted a 1:11.40, which put him seventh. That's Q3, then! It was all going well so far.

The effect that this kind of weather and this kind of situation has on the F1 grid is pretty fascinating. The order of things is threatened so all of the rich bastards at the top of the pile start getting nervous and all the poor people start rubbing their hands. Sure enough, as Q3 started, the sky went dark and as everybody started wetting their shorts we had a decision to make. Either we sent Kevin out on softs or we sent him out on intermediates.

Because of where we're positioned on the pit lane, we're closest to the pit lane exit and just for once we were able to make that work for us. Put him on softs, send him out first and hope he can post a good lap before it rains. That was the plan. I remember watching the queue form behind Kevin from the pit wall. They were there for over a minute and the noise was foking deafening. All the time the tyre temperatures were falling, though. 'Just let them go!' I remember saying. A few seconds later the lights changed to green. 'OK, you've got a clear track, mate,' said Mark, who is Kevin's new engineer. And that was it. He was gone. We'd taken our chance and we just had to pray it worked.

Kevin ended up posting a time of 1:11.674.

'Where does that put me?' he asked Mark.

'You're P1, mate.'

'You're kidding, you're fucking kidding me,' he replied.

I thought to myself, *Don't get foking carried away, you guys! We might get a black and orange flag for something!*

'Ayao,' I said. 'Get off your stool and do a foking rain dance. Come on, quick! We need some help here.'

'I don't need to,' he said. 'Look.'

I turned around on my stool and saw heavy rain falling on to the pit lane. Talk about timing! The drivers had just started their second laps and Perez had gone fast in the first sector. Jeezoz Christ. Thank fok for that!

'How long is it due to rain for?' I asked Pete.

'About half an hour,' he said.

The next thing I remember is hearing that George Russell had come off at Turn 4. 'It's a red flag,' said Pete. 'And there's eight minutes and ten seconds left. Kevin's still P1.' For a few seconds the three of us sat there on the pit wall trying to figure out what could go wrong. We've had our fingers burnt so many foking times this season and we honestly expected it to be taken away from us. After about a minute nobody had thought of anything so it had actually happened. We were on pole!

There, that's what happened as I remember it.

The only member of the team who hasn't really been able to share in the celebrations is Mick. He's been very gracious about it and has congratulated everyone but he's obviously hurting. At the end of the day, though, the grid doesn't lie and he's at the opposite end to Kevin. It's easy for me to say this but he has to try and take something positive away from this. Something that will help him improve and become stronger. If he was the same age and had the same experience as Kevin it would be a lot harder for him to take for sure. He's young, though, and he'll live to fight another day. The best thing he can do is come back tomorrow and drive the race of his foking life. I hope he does.

OK, so what does this mean to team Haas? First of all, it's obviously a reward for all the hard work the guys put in week in week out. It doesn't matter whether we qualify twentieth or on pole just like we did today, these guys do the same things and with the same amount of effort and attention to detail every single time. In the vast majority of cases, what happens after that is out of their hands so to maintain that level of quality regardless is proof of their professionalism.

Some people might think that working for an F1 team is just a job. After all, we get paid the same as everyone else and it helps to keep a roof over our heads. That isn't why we're here, though. The people who work for us want to be part of a team and they want that team to progress as much as possible and achieve things. This is another box we have ticked together. We've had points, fastest laps, and now we have a pole position to our name. Next stop, a podium.

It might sound a bit foking corny I suppose, but working for a small team like Haas *is* a way of life and everything that happens in the team matters, whether it be a puncture that takes you out of the points or a black and orange flag that makes you want to kill somebody and almost gives you a foking heart attack, or a pole position like today. Because you live it day to day you feel everything very acutely and Kevin's pole position is a reminder of why we do it. A timely reminder.

Poles are routine for teams like Red Bull and Mercedes so, although you have more success, it becomes expected. What was it I said earlier? When there are always biscuits in the foking tin, where is the fun in biscuits? I'm not saying it doesn't mean anything when the big teams get a pole or win a race, but surely the level of success a team member feels will be measured partly by contribution and so the smaller the team the more invested you're going to feel. Not just when things are good, but when they are shit, too. This is just a theory, by the way, so if you work for Red Bull or Mercedes, don't come down the paddock and start kicking my ass. Or if you do, bring me one of your foking trophies!

When I said earlier that being on pole means you exist in a different world, as well as creating new dreams, it also creates a new reaction. Not just from the fans and the people around you but from the sport in general. When it was confirmed that Kevin was on pole position, there was a roar from the crowd, the like of which you only usually hear when a home driver wins a Grand Prix or a driver wins a championship. It was just unbelievable. I was on the pit wall with Pete and Ayao and when I turned around the entire foking grandstand were on their feet going crazy. Going crazy for us. For Haas. I then looked down at the garage and of course they were all going even crazier. I later saw that some of them were in tears, you know. For a few moments I just sat there and took it all in. I obviously have no idea whether something like this will ever happen again for Haas. I hope it does but I wanted to savour every second. When I eventually did cross the pit lane to join the guys in the garage, the roar from the grandstands started again and I remember lifting my hands in the air to celebrate with them. 'Yes, we foking did it!'

The next two hours are a bit of a haze, to be honest. I must have been congratulated by at least a thousand different people and everybody who congratulated me meant it and was happy for us.

Every fan, every journalist, every driver and every team principal. This isn't just good for Haas, you know. It's good for Formula 1. The entire sport celebrating something together. I can't think of another sport where something like that would happen.

When Kevin arrived back in the garage, he jumped out of the car, stood on top of it, jumped up and down and went absolutely foking bananas. And then the hugging started. Every member of the team was in there and everybody wanted to congratulate him. I'm so happy for the little Viking. He has been through every emotion known to man in his F1 career so far, except the emotion you feel when you get a pole or win a race. Being first, in other words. It's the same for all of us really. We're all feeling a version of that at the moment and it's pretty incredible. I've watched the video of him getting out of the car about ten times already. If he's damaged the foking car I will kick his tiny arse from here to Copenhagen. Viking or no Viking!

After finally managing to remove Kevin from my leg, Stuart led me to the paddock to talk to the media. And not just some of the media. All of the media! Every journalist and TV station in the circuit wanted to talk to me and Kevin. And do you know what, we wanted to talk to them! When I emerged from the garage to start speaking to them, the cheering started again, except the people cheering were right there in front of me. Fok me, it was loud! Everybody likes seeing smiling faces but when they're smiling because of something your team has achieved . . . Well, that is extra special.

If I could bottle one part of today, it would be the moment I realized that we had pole. On the pit lane we obviously have all the lap times in front of us and I remember seeing Kevin's name at the top. And then the rain started. 'Ayao, Pete,' I said. 'That's it. We've got pole. We've got foking pole!' I think they were both speechless, you know. They just sat there smiling their heads off. It was a moment that I have been dreaming about and fighting for since I started the team with Gene all those years ago and it will stay with me for the rest of my life. As I said to Kevin after the session, whatever happens tomorrow, nobody will ever be able to take this away from us.

Saturday, 12 November 2022 – Interlagos, São Paulo, Brazil

10 a.m.

I'm not sure if the population of Brazil has suddenly been ordered to watch *Drive to Survive*, but the number of people who want a photo today compared to yesterday and Thursday is just ridiculous. I must have posed for about three hundred already. It started when I was about half a mile away from Interlagos at some traffic lights. Somebody shouted 'GUENTHEEEEEER!' and, before I knew it, my car was surrounded again by mad foking Brazilian people. Fortunately, a couple of policemen came and saved me, otherwise I'd still be there.

When I finally got to the entrance, it started again. The traffic was shit, though, so there was nowhere to go. I don't speak Portuguese so I actually have no idea what they were saying to me, apart from 'GUENTHEEEEEER!' I hope it was complimentary. I didn't miss anybody out, though. Everybody got a selfie, whether they wanted one or not! I have to say it's a lot quicker than signing foking autographs.

The bit that took me the most amount of time was when I got to the paddock entrance. After climbing all those stairs I was foked and was in no position to fight. 'GUENTHEEEEEEER! WE LOVE YOOOOOOU!' I must have done a couple of hundred at least there and in the end one of the security guards had to pull me through the barriers and push me towards the turnstiles.

Everybody is still buzzing from yesterday. We have the Money-Gram International guys with us again and they're having a great weekend. Since we announced the sponsorship just before Austin, we've scored double points and have got a pole position so it's been a good start! They're obviously a lucky omen for us so they should come to every race. We're all going out for dinner later to a restaurant called Barbacoa, which I've been to before and is excellent. They do the best meat in the world there and in preparation I haven't eaten any meat for the last two days. I have no idea why I am telling you this. It's just boring bullshit! We've got a race in a few hours.

1 p.m.

The messages of congratulation are continuing to arrive, both in person and on my phone. It's only now that we've been able to gauge the true response to what happened and to be honest I think it's taken us by surprise. Everybody seems to be a Haas fan at the moment and I suppose it's because most people love it when an underdog does well. I'm good with that. According to Stuart, social media has been off the scale. One of yesterday's Instagram posts has had over a million likes. That means nothing at all to me but Stuart and his team say that's very good. This kind of coverage and engagement obviously presents an opportunity for us and they've been working their absolute asses off to make sure we grab it. It's an exciting time for Haas F1.

So, what am I expecting from the race today? Well, I'm very glad I asked myself that question because it's relevant to the race and to what's been going on generally. Despite us being pleased about the pole, our reaction has had to be quite businesslike. We're only halfway through the weekend and have a sprint race and a Grand Prix in front of us. The job is only half done, in other words. It's the same with the race. As amazing as the qualifying was, unless we come home with some points it won't be as good. I know Kevin will be disappointed. We're also realistic with regards to our chances. An element of luck played a part in us getting pole and nobody thinks for one moment that we're going to now storm to victory. It'd be absolutely foking amazing if we did, but it isn't very likely. There are probably six cars that are faster than ours and another four that are just as fast. A podium is always possible but there would have to be some surprises. To be in the points. That's our goal.

6 p.m.

Kevin finished eighth and Mick twelfth so that's a point in the bag and a much-improved performance by Mick. It's funny – although we've been managing our expectations pretty successfully since yesterday, there was a period at the very start of the race when I began to dream a bit. Kevin had a very good start and after a lap he was about a second ahead of Verstappen, which is when I started to dream. There were a couple of battles going on behind Kevin but, as things started to settle down, he was gradually caught and moved down the field. Kevin did exactly what he had to, though, and that extra point puts us two ahead of AlphaTauri, with two races to go.

Mick made up eight places and I think that's helped him scrape back a little bit of confidence. Yesterday hit him really hard and so to come back like that and drive a good race shows character. It also puts him in a good place for tomorrow.

Sunday, 13 November 2022 – Interlagos, São Paulo, Brazil

8 a.m.

I don't know where it came from but all of a sudden I have a foking camel cough. I feel like shit and sound like I'm dying! I went out last night with the MoneyGram guys and felt fine then. I only had two small glasses of wine. Although I did overeat a bit. Anyway, that has nothing to do with this foking cough. The only thing I can put it down to is a sharp change of temperature yesterday. If it carries on I might have to see the F1 doctor. Poor Guenther.

1 p.m.

The doctor says that it could be the change in temperature and that I should drink plenty of water. What, you mean as opposed to whisky? At least he didn't tell me to try and get plenty of rest. I'll sleep when I'm dead!

9.30 p.m.

I've been too foking angry to write anything until now. Or at least anything that won't get me into some serious shit with the FIA. I'm almost beyond caring now. I admire their consistency, though. If we were as consistently good as they are poor we'd be a top-three team right now. Unfortunately, we're the team that seems to suffer most from their mistakes and I'm up to here with it.

So, let me tell you about the latest FIA debacle. Kevin got taken out by Ricciardo at Turn 8 on the first lap of the race. That in itself is obviously shit but then he was left there in the middle of the race track for over two foking hours! Nobody came to get him. When you are taken out like that so early in a race all you want to do is get back to the garage, calm down and be with your team. Instead, Kevin was left in the middle of Interlagos with a few thousand fans for company. Imagine if that had been Lewis Hamilton. He'd have been torn to foking pieces! And Kevin was pissed off, you know. Seriously pissed off. He's normally quite a placid guy but not today. He wanted blood! Ricciardo's and the FIA's, preferably.

Speaking of McLaren, do you know what else happened? Apparently they tried to blame Kevin in front of the foking stewards! I received a text message from their chief executive officer, Zak Brown, apologizing earlier, yet his team have tried to blame us? He needs to have a word with them and I sent him a text back earlier saying so. Mick finished thirteenth, by the way, so all that promise went to shit in the end. At least AlphaTauri didn't score any points. At one stage it looked like Gasly might, but then he got a five-second penalty for speeding in the pit lane. One more race to go.

On the drive back to the hotel, I called Ronan Morgan at the

FIA and then Mohammed Ben Sulayem. Ronan was Mohammed's co-driver in his rally days and he's now his right-hand man. I didn't hold back with either of them. At the end of the day, the mistakes that have been made by the employees of the FIA, in addition to leaving a driver stuck in the middle of a foking race track for two hours among thousands of fans, could potentially have cost Haas millions of dollars. Also, if AlphaTauri score points at the last race and we don't it could end up costing us another ten million dollars. It's that serious, you know. The black and orange flag debacle has been a joke all season and every time I have complained to them not one person at the FIA has been able to defend what has happened. And it was the same earlier. Neither Ronan nor Mohammed were able to defend the actions of their employees in any instance I mentioned. So what does that tell you? This is Formula 1, for fok's sake. Not Formula Regional! Anyway, I've told them that I want to sit down with them both in Abu Dhabi and sort something out because I am not leaving it like this. No way.

Let me give you an example of how inconsistent things have become. You remember the penalty Alonso received when we complained about his mirror? The one that moved us up a place and gave us extra points? Well, Alonso and Alpine appealed that decision and the stewards capitulated. It was foking unbelievable. When we made the original complaint we understood that the race director had said we had one hour to submit it officially, and so we did. The rules state that complaints should be submitted within thirty minutes from the end of the race (which was the basis of Alpine's appeal) but when the race director says you have an hour you take his word for it. What really annoys me is that when we had a meeting about it the race director denied saying what we thought he'd said and wouldn't even apologize. And he's the race director, for fok's sake! If he'd said, 'OK, guys, look, I'm afraid I foked up,' we'd have been OK about it. Annoyed still, but ultimately OK. He didn't, though.

OK, the car comes in thirty minutes to take me to the airport for the flight to Dubai (we get a coach from there to Abu Dhabi) so I'd better get my shit together. I'm still coughing, by the way.

Boy, what a weekend. It really has had everything.

SEASON FINALE

Tuesday, 15 November 2022 – Yas Marina Circuit, Yas Island, Abu Dhabi

5 p.m.

Do you know, I think I might have had something similar to what I had a few weeks ago. Not anywhere near as bad, of course. When I got on the plane I slept all the way to Dubai (ten hours), then all the way to Abu Dhabi and then slept for another six hours when I got to the hotel. I feel OK now but I've definitely had some kind of virus.

The big news from Haas is that Gene and I have decided not to retain Mick for next season and beyond. Instead, we're going with Nico Hülkenberg. I know this is going to be a very unpopular decision in some quarters but, as I said before, I have to do what I think is right for the team and what I think is right for the team at the moment – especially if we want to move forward next season – is a more experienced driver. I think Mick already knows it's going to happen and, as far as I know, he is already being lined up to be a reserve driver at Mercedes next year. In the absence of any other full-time F1 seats, it'll be the next best thing. I hope it happens for him.

Despite what a lot of people might think, I like Mick and I stand by our claim that we have supported him throughout his time with Haas. Could we have done things better sometimes? Of course we could. Nobody is perfect. I sound like I'm making a foking statement here! It's true, though, and I hope that Mick agrees that we have supported him. It's been a difficult situation for everybody. It's also been quite political sometimes (especially towards the end) and the media haven't helped matters.

The fallout when we announce it will probably be a bit shit but I'm ready for it. I've had to make far more difficult decisions than this in my career and I've had to put up with far more shit than anything the German media or Mick's uncle could ever throw at me. I think it's my persona of being a joker that gives people the wrong impression. It makes them think there's nothing underneath and that I'm just some kind of clown. I've been working in motorsport for over thirty years and I know a foking thing or two. I've said it before and I'll say it again. This team is my life and when it comes to protecting it I'm willing to take on all-comers. Don't make me angry, that's all. You wouldn't like me when I'm angry!

Talking of Hulks.

The discussions with Nico started a little while ago. One of the things that's been missing for us, I think, is somebody who has experience of driving for more than one team and Nico obviously provides that. He's also a good qualifier and very solid racer. There'll be people saying he's too old and hasn't had a full-time drive for three years but I couldn't give a shit. If I didn't think he was up to it I wouldn't have had a conversation with him. I'll say this for him, he's been very keen. Being a reserve driver, like Nico has been with Aston Martin, can give you a very good life. The money's usually good (unless you work for me and Gene) and at the end of the day you don't have to do a great deal. I wanted to see just how keen Nico was to go full time again and so, after our initial conversation, I didn't call him back. I thought, *If he wants this badly enough, he'll get in touch with me.* Over the next week he must have called me about ten times and it got to a point where I almost told him to fok off. OK, I get the message!

Anyway, I'm going back to the team hotel to get an early night. Tomorrow's going to be a big one.

Wednesday, 16 November 2022 – Yas Marina Circuit, Yas Island, Abu Dhabi

5 p.m.

I had my chat with Mick earlier. It wasn't an easy conversation but I explained our position and did my best to help him understand our reasons for not retaining him. He's actually quite mature for a young guy and he was calm and respectful throughout. One of the reasons he feels so disappointed, I think, is because he feels like he's been making some progress this season. Unfortunately, that progress hasn't been quick or consistent enough for us to take a chance on retaining him. I wish I could say differently, but I can't. Some people will think it's just the crashes that have made our decision but that isn't true. Sure, it's been a factor. That too comes from a lack of experience and, unfortunately, we just cannot afford to give

Mick the time he needs to progress. I had to have a similar conversation with Kevin and Romain a couple of years ago but for different reasons. That too wasn't easy but it had to be done.

We're going to go public with it tomorrow so now everybody has time to prepare.

Thursday, 17 November 2022 – Yas Marina Circuit, Yas Island, Abu Dhabi

4 p.m.

Well, I'm finally back to my old self again, and just in foking time. When Stuart pulled the trigger earlier it suddenly went crazy and I've been doing interviews ever since. Apparently social media is not a good place to visit at the moment, especially if you are a Haas supporter or Guenther Steiner. Some people seem to have forgotten that Mick's contract has come to an end and our only crime is that we have decided not to extend it. We don't owe Mick a drive for next year. In our opinion, he hasn't done enough to merit us offering him a new contract and so that's the end of it. Things change and you move on.

Over the past couple of years, things have got a lot worse on social media with regards to abuse. I suppose that could also be down to *Drive to Survive*. At least partly. Some of the shit we had thrown at us after the Uralkali and Mazepin episode was just horrendous. Fortunately, I don't look at it very often but, according to Stuart and his team, it's been getting worse for a while now. In my opinion, real F1 fans do not behave in this manner. Sure, they like to complain sometimes. We all do when things aren't going our way. But they do it respectfully. I hope I'm right about this because if real F1 fans are becoming abusive like these keyboard idiots it will be a sad day for the sport.

With regards to the interviews so far, most people seem to understand our reasons. Unfortunately, that doesn't make a good story, though. What makes a good story is Mick being badly done to by nasty, horrible old Guenther. That's OK. I know how the game works. If Mick's surname wasn't Schumacher nobody would care about this. It is, though, and so they do. Don't misunderstand me. A lot of pressure comes with having a seven-time world champion for a father and on the whole I think Mick has handled it very well indeed. His entourage and supporters, perhaps not so much. It's also not his fault that we're getting attacked like we are. Once again, that comes with the surname. If you're a Schumacher, people will support you, and support you no matter what.

I find driver managers in general to be quite frustrating these days. I might get into troble for saying this, but it's true. This is only an opinion, though, so don't start getting your foking knickers in a twist! In my experience, as opposed to actually managing the drivers and doing what is best for them, they often tend to do what is asked of them. And there's a difference. A big difference. It's almost as if they have forgotten how to do that side of the job. Or maybe they just don't want to? Perhaps they're scared.

I've discussed this with a friend of mine in F1, who I cannot name but who speaks much better English than I do, and they call it the path of least resistance. Let me give you an example. Not a specific one. A general one. If a driver says they want some overalls that are the wrong colour for the team, some of these people would sooner ask the team to change their colours than the driver.

I don't want to sound like an old fart who says that it was always better in the olden days (it foking wasn't), but when I first started in Formula 1, the managers were like an extension of the drivers they represented. And, as opposed to being just yes men, they always had their best interests at heart. They also did not mind bollocking their drivers and saying no to them. These days, if a driver isn't happy, a manager will come to you and say, 'My driver isn't happy,' whereas what they used to do is find out why they weren't happy and fix it. Sometimes that would involve the team, and sometimes it wouldn't. The point being, it was their responsibility to make the driver happy again, not the team's.

Look, I get that things have changed these days and that you can no longer have just one person looking after a driver. What you can have, though, are people, or a team of people, who are actually going to manage the driver, as opposed to just saying yes all the time.

The number of people who are involved in a driver's career has also had an impact on the relationship they have with a team principal. You remember Michael Schumacher and Jean Todt, right? That kind of dynamic does not exist any more and only existed in the first place because he was so foking successful. I think that when Mick came to Haas, some people hoped that history might repeat itself. Times have moved on but there were also several world championships missing.

Another thing that the managers these days are very good at is blaming the team when something goes wrong. They're experts at it and, once again, it is the path of least resistance. It also makes the driver think that they are looking after their best interests. If I fail at something I cannot blame Ayao or Pete or Gene Haas. I blame me and I have to take full responsibility.

What makes me laugh sometimes is the language that a driver will use. Again, I have had a lot of discussions about this over the years with my colleagues and fellow team principals. If a driver has a good result they'll use 'I' and 'me' in the interviews afterwards. And, if they have a shit result, they will use 'us' and 'we'. I'm not sure if that is taught or if it's natural, but it's clever. 'Who is this foking "we"?' I remember a team principal saying to me one day. 'Is there somebody else in the foking car?'

I've been hearing and reading a lot of quotes in the media about Mick deserving a seat in F1. What a load of absolute foking bullshit! Nobody deserves a seat in F1, just as nobody is entitled to a seat in F1. If you're good enough and you fight for it, you might be lucky enough to get one, but you do not deserve that seat. If you get it, you have simply earned that seat.

We as a team want to do something different, so why do we have to ask anyone for permission? The answer is we don't, and the sooner people wake up and realize that all we have done is simply exercise our right not to renew a contract, the better.

I haven't heard anything from Mick's uncle yet. A couple of days ago he had another go about me to the press. This time he said that, and I quote, 'I believe that Guenther Steiner cannot deal with the fact that someone else is the focus at Haas. He's very, very happy to be the one front and centre.' Jeezoz Christ! What does he think? That I was 'invented' for *Drive to Survive*? It's not my fault that people know who I am. I did not do it on purpose. Honestly, how can people be so stupid?

One of the journalists I spoke to earlier repeated what Mick's uncle had said in the press and then offered me a right of reply. What a waste of everyone's time. It's foking unbelievable. I couldn't give a shit whether the focus of attention is on Mick, me or Martin Brundle's foking armpits. All I care about is whether or not the people who drive for Haas can deliver for us on the track. Seriously, if that's the best that Mick's uncle can come up with he should get himself a paper round because he's shit at making headlines. This has nothing whatsoever to do with Mick, by the way. Me and him are OK.

With regards to Nico, some of the journalists I spoke to earlier brought up an argument that he and Kevin had back in 2017. In Hungary, I think. If they hated each other I might have thought twice about bringing Nico in but, as far as I know, they're good now. It doesn't matter if they're not the best of friends, as long as they're respectful to each other and respectful to the team. They're drivers who I think are on a similar level, ability-wise, and have a similar amount of experience. It's going to be interesting to see what happens next year. I can't wait.

Friday, 18 November 2022 – Yas Marina Circuit, Yas Island, Abu Dhabi

8 a.m.

There is an F1 Commission meeting today. If you could see my face now you would know that I can hardly contain my excitement. I'm just not in the mood for this at the moment. Do you know how long they last? About four foking hours! Four hours that you never get back. I am trying to find a positive spin to put on it, apart from a free bottle of water and a comfy chair, but I can't think of one. Not at the moment.

Shit, there's a lot of people here this weekend. I know I've said it before in the book many times now but because of all the stuff that's happened with Mick and Nico I haven't really noticed them until now. If things go the same way as they have over the past couple of years, I can see every Grand Prix being as busy and intense as the likes of Mexico one day. They are going to have to think about these things, though. As much as we enjoy interacting with fans – and we do, believe me – it cannot be to the detriment of the job that we're here to do. I can't afford to take twenty minutes out of my schedule every time I leave the garage or my office. Perhaps I should try and bring it up at the F1 Commission meeting. It might help to keep me awake!

9.30 a.m.

The fallout from yesterday is still ongoing but at least some of the questions have been interesting. One journalist asked me if I thought Mick had it in him to become an above average F1 driver, and I said yes. I then repeated my explanation that next season we need to try and take two steps forward and by waiting for Mick to mature and improve we could be taking two steps back. And what if it never happens? I didn't tell the journalist this but I also think that Kevin has been in a bit of a comfort zone. You know, *Things are OK here. I'm fine.* I'm not saying that he's been driving down exactly, but at the same time he hasn't been looking over his shoulder enough. If you discount the retirements and the Saudi Arabia Grand Prix where Mick didn't start, Mick has actually finished ahead of Kevin in twelve out of seventeen races. Marginally so, but ahead. What he hasn't done, however, is qualify anywhere near as well as Kevin or score anywhere near as many points. And you cannot discount the crashes, I'm afraid. I wish I could, but I can't. Grosjean could be quite similar sometimes. You know, a bit unpredictable. He was fast, though, and on his day could take on all-comers. Mick has the first bit right but not the second. At the moment. I think the cost of Mick's crashes this year totalled over $2 million. But we're the bad guys.

I might be the team principal, but the people who push our team really are the drivers. We give them the tools to do so, but they are the ones who score the points, keep us competitive (hopefully) and make us want to come back and do it again. Without them we are nothing and Gene and I owe it to everyone at Haas F1 to make sure we have in place the best drivers we can afford to make that happen. Nico will push Kevin and vice versa. Mark my words. You will see a different Kevin Magnussen next season.

Anyway, I'd better drag myself to this meeting.

9 p.m.

Wow! I managed to stay awake for the whole meeting. And I didn't have to take drugs or stick matchsticks in my eyes. First the FIA president, Mohammed Ben Sulayem, made a speech congratulating everybody on a great season, then Stefano made a speech congratulating everybody on a great season, and then we got to business. Subjects covered today were the tyre blanket strategy (the FIA want to remove tyre blankets from 2024 but a decision has been delayed due to driver concerns and feedback); DRS activation after the start, re-start, or safety car (the Commission approved a proposal to assess a method of keeping the field closer together and encouraging closer racing by bringing the activation of the DRS forward by one lap at the start of a race or sprint session, or following a safety car re-start); Parc Fermé for events that include a sprint session (we discussed ways of simplifying everything, basically); accident damage allowance (they have now simplified the system in place to deal with the impact of accident damage during a sprint); and technical and financial regulations. This was about the introduction of stronger roll hoops, which will appear in 2024. If you'd had the season I've just had, could you stay awake for all of that? Of course you couldn't. I must be super foking human.

Because of all that excitement, I almost forgot to tell you about Free Practice earlier. It wasn't the easiest session today. Kevin sustained damage to his floor in both sessions and the pace wasn't where it should have been. Pietro did a good job in FP1. He had a bit of a shit one last time out but managed twenty-six laps this time and ended up P15 on the time sheet. He's also doing some testing with Nico here next Tuesday, so it will help him prepare for that. Mick came back in for FP2 and just needed a few laps to get used to the track. He too suffered with a lack of pace so it's something we're working on.

OK, fok off. I'm tired.

Sunday, 20 November 2022 – Yas Marina Circuit, Yas Island, Abu Dhabi

6 p.m.

Well, we managed to secure eighth in the Constructors'. That's been our goal for a while now so to get there is a nice little victory to end the season on. It wasn't a nice experience, though. I'll be honest with you. The thing I found most disturbing was the fact that we had no influence at all over what happened. Not being able to defend yourself with your own weapons is a pretty shit feeling. All we could do was sit there on the pit wall and in the garage and hope that a crash didn't occur that would let Tsunoda into the points. If you want hope, you go to church! I don't want to be in that situation again.

10 p.m.

I had a choice of flying out of Abu Dhabi tomorrow morning at 3 a.m. or flying out tomorrow mid-afternoon and I chose tomorrow morning. It's not the healthiest thing to do but I just want to get back home as quickly as possible. Get this, though. I'm changing at foking Frankfurt! Can you believe that? I'd better watch my back! I'm public enemy number one in Germany at the moment.

Anyway. I'm going to call Gertie and Greta now. Then I'm going to have a shower, pack my stuff and work out what I'm going to do in Frankfurt.

Tuesday, 22 November 2022 – Steiner Ranch, North Carolina, USA

11 a.m.

This will make you laugh. You might have thought I was joking when I said that I might get some shit at Frankfurt Airport. I actually wasn't. Michael's obviously like a god over there and, being Michael's son, Mick is thought of in the same kind of way. Not re-hiring him was like not re-hiring Jeezoz in their eyes!

I made sure I didn't travel in branded clothing and kept my sunglasses on. Even so, about thirty seconds after stepping off the plane I was approached by a group of four men. Four quite stern-looking men. I thought, *Jeezoz Christ, here we go. Another load of people who think I owe Mick a foking living.* Do you know what they did? They held their hands out and congratulated me on the pole position in Brazil! No mention of Mick at all.

I told the publishers that I'm going to try and finish the book today and they asked me if I could write what I think will happen to Haas and Formula 1 in the next five years. I said, 'You have to be foking joking, right? Everything I have ever predicted before has turned out to be wrong. Trust me, it will all come to shit.' 'Just write it down,' they said.

OK, let's start with F1. I actually don't think a great deal will have changed, to be honest with you. The need for the sport to sit back, refine what's happened and manage the growth is obvious and that isn't going to happen overnight. There might be another race or two on the calendar and one or two of the teams might have changed hands. Then again, although I said that the easiest way to get into F1 was to buy a team, what if nobody is selling? If that's the case I think F1 will start receiving licence applications from some very high-profile people soon, with a lot of money and a lot of influence and resources behind them. You heard it here first. Or read it.

So where do I think Haas will be in five years' time? Well, I probably will have been cryogenically frozen to stop me from talking bullshit all the time. Conor McGregor will have bought the team from Gene for a billion dollars and 'our cars will be exceptional'. Mick's uncle will be the team principal, Mick and Nikita will be the drivers, Fred will be the mascot, Mattia will be supplying the wine and Toto will be in charge of sponsorship. That's the ideal scenario, at least from the press's point of view.

Now let me tell you what I really think.

For a start, I think the future is very bright for Haas (and for every team on the grid, come to think of it) and in five years' time I think that we will be a successful and profitable organization. If that is the case, why would you not keep on doing it? When I first started talking to different people about setting up an F1 team all those years ago, Formula 1 was basically a plutocracy, which is why there was always so much change going on in the midfield and at the back of the grid. Unless you were at the very top of the pile, you were almost guaranteed to lose a shitload of money and so the passion with these teams used to run out pretty soon. You remember what I said about nobody wanting to buy Manor for a pound? It was a closed shop. The big teams got richer and the poor teams either changed hands or went to shit. Even if we'd had the infrastructure in place to be completely self-sufficient, we would have been lucky to last two seasons. So, for us at least, the only way was with Ferrari.

Even though we have been in existence now almost for ten years (eight years on the grid), we are still the youngest team in F1, which means things have gone full circle. Unlike ten or twenty years ago, we have ten teams that look like they will survive long term. Survival is not the goal, though. Not for me. I would not do that in my life. I say it again, I am here to compete.

When Gene and I started working together, we had a five-year plan, but that was really just to get into Formula 1 and then stabilize. Once you are here, five-year plans cannot exist because things are evolving all the time. Look at Formula 1 today. Who could have guessed five years ago that by 2023 even one of the smaller teams could be worth half a billion dollars? Somebody would have laughed in your foking face if you'd suggested that five years ago. Ten years ago they would have reported you to the authorities. And remember, in the middle of that five years has been a pandemic! How the fok did it all happen?

One day I will be too old and will have to stop doing this and my dream is that when I walk away Haas F1 will be a success both on and off the track. But again, who says I will last that long? Gene could turn around tomorrow and decide that he wants to sell the team and if that happens I could be looking for another job. And I wouldn't blame him, you know. He's the one who took the risk. Why not take the reward?

I don't actually want him to, so if you're reading this, Gene, do not do anything stupid! If Haas F1 is worth half a billion now, imagine how much it could be worth when I've finished with it.

Actually, don't imagine that. To be serious, though, when I walk away or when Gene sells the team, we will want to make sure that everybody who is employed by Haas F1 is not only part of something successful, but has job security.

If Ferrari gave us three years' notice on our contract tomorrow and said that after that we would be on our own, I wouldn't be very happy but I think we would survive. Obviously, we would have to be very careful. The two major changes would be making our own gearboxes and suspension. It doesn't sound like very much, but believe me, it is. I would like to stay with Ferrari, at least in the medium term. If it isn't bust, don't try and fix it.

Regardless of how much the team might be worth right now, we have just had to weather a pandemic, a season that had us going backwards, and another season that had more drama in it than a thousand episodes of a foking Mexican soap opera. Haas, like Formula 1, need to stabilize and try and have a couple of seasons where we can just concentrate on trying to improve on the track. We just want to go racing, for fok's sake! There will obviously be some dramas here and there, but hopefully not pandemic-shaped ones, or ones that involve large countries beginning with the letter R.

Anyway, I think it's time I wrapped this book up. You've probably had enough of me complaining by now. I foking have! It was completely unintentional but I couldn't have picked a better season to do a diary like this. The fact that we were coming back with a new car after what had been a pretty shit couple of years was always going to make it interesting. I could never have dreamed it would be as dramatic as this, though. You lucky fokers! I should be charging you double.

OK, let me recap what has happened.

We had to get rid of a driver and a title sponsor before the season had even started (which was absolutely foking marvellous), hired a miniature Viking who we already knew quite well to take the driver's place, scored points in our first race, scored points in our second race, fell out with the German media on lots of occasions, qualified P4 in the sprint race at Imola, made friends with Miami, met Conor McGregor in Monaco, who is a foking Haas fan, scored double points at two races on the bounce, including Silverstone, collected enough black and orange flags to cover the United States of America, got very, very angry with the FIA, drank a bit of wine with Mattia from Ferrari, took the piss out of Fred from Alfa, smoked cigars with a friend, got an amazing new title sponsor, scored points at our home Grand Prix, got zip in foking Mexico again, helped to launch the Las Vegas Grand Prix, taught Gordon Ramsay how to swear, was driven to the brink in Japan and got a pole position at Interlagos! I've probably missed out lots of things but even that is a good list.

There has been a lot of talk about my management style, especially since we decided not to renew Mick's contract. I certainly get asked about it a lot now. People mainly want to know if I think it has changed over the years and I don't think it has. I've always been quite direct and I've always encouraged people to tell me the truth and not just something I want to hear. Whatever you think, I'm too old to start changing now. At least to any great degree. Can you imagine me going on a foking management course? No, neither can I.

We also lost one of our own this year, which puts everything into perspective. Harvey loved Formula 1 as much as anyone I have ever known and the fact that his team have made progress this year would make him a really happy guy. Knowing that makes everyone at Haas F1 smile. This book is dedicated to him, by the way. It's for our friend and colleague, Harvey Cook.

OK, I have to go now. I have two hours off before I start preparing for the new season. You think I'm kidding? Come on. You know by now that I never stop. It's been good fun and I hope you have enjoyed it. And who knows? Maybe I'll do it again. It's not really hard work. I just write stuff down occasionally, send it to the publishers who then send it to the lawyers who then shit themselves, and then somebody corrects the spelling, tidies it up a bit and tries to take out all of the swearing. No foking chance!

Take care, you bunch of wankers. See you on the other side!

ACKNOWLEDGEMENTS

First, I would like to thank James Hogg for approaching me with the idea of writing this book and for helping me put it all together. The things people will do to get out of Leeds for a few weeks, eh, Hoggy? Anyway, thank you. It's been good having you around.

I would also like to thank Tim Bates from Peters Fraser + Dunlop, and Henry Vines from Transworld, who have also worked very hard on my behalf. It's much appreciated.

Without *Drive to Survive* I would never have been asked to write a book in the first place. Nor would there be thousands of T-shirts in existence with my ugly face on them. Have you seen what you've done, guys? They're good people and we always have a lot of fun, so thank you.

When you work as part of a team, you're only as good as the people you have around you, and at Haas we're lucky to have some of the best. This book isn't just about my year, it's about *our* year and the adventure we went on together. You never stopped fighting and I'll never forget that. Here's to us.

There is one person at Haas who has had to keep me in check throughout this entire process and that's our communications director, Stuart Morrison. I couldn't have done this without you Stuart, so thank you.

Next up, I would like to pay tribute to my boss, Gene Haas, for giving me permission to write this book and for not asking to read it first! Thank you, Gene.

Last but not least, I would like to say a big heartfelt thank you to my wife, Gertie, and my daughter, Greta. As much as I love what I do for a living, it's knowing that you're both there that keeps me going.

PICTURE ACKNOWLEDGEMENTS

All photographs supplied courtesy of Haas F1 Team, with the following exceptions.

Every effort has been made to contact and acknowledge copyright holders, but the publishers apologize for any errors or omissions in this respect and invite corrections for future editions.

Page 1: Safari Rally, 1992 © McKlein Photography; with Colin McRae in Monaco, 1999 © Sutton Images; with Eddie Irvine at the French Grand Prix, 2002 © Bryn Lennon/Getty Images; with Niki Lauda at Silverstone, 2002 © Mark Thompson/Getty Images.

Page 2: at the Haas F1 press conference in North Carolina, 2014 © Jared C. Tilton/Getty Images.

Page 5: with Conor McGregor in Monaco, 2022 © Arthur Thill ATP Images/Getty Images.

ABOUT THE AUTHOR

Guenther Steiner is an Italian motorsport engineer and team manager. He is the current team principal of the Haas F1 Team, and the previous managing director of Jaguar Racing and technical operations director of its subsequent incarnation, Red Bull Racing. In 2014, Guenther persuaded Gene Haas, owner of Haas Automation and NASCAR championship-winning team Stewart-Haas Racing, to enter Formula 1.

With their entry in the 2016 season, Haas became the first American constructor to compete in F1 in thirty years. The team took eight points at the 2016 Australian Grand Prix with a sixth-place finish, becoming the first American entry, and the first constructor overall since Toyota Racing in 2002, to score in their debut race. Steiner is also a prominent figure in the cast of the successful Netflix series *Drive to Survive*.